HANDBOOK OF ELECTRONIC MATERIALS
Volume 4

HANDBOOK OF ELECTRONIC MATERIALS

Compiled by:

ELECTRONIC PROPERTIES INFORMATION CENTER

Hughes Aircraft Company
Culver City, California

Sponsored by:

AIR FORCE MATERIALS LABORATORY

Air Force Systems Command
Wright Patterson Air Force Base, Ohio

HANDBOOK OF ELECTRONIC MATERIALS
Volume 4

Niobium Alloys and Compounds

M. Neuberger, D. L. Grigsby, and W. H. Veazie, Jr.

Electronic Properties Information Center
Hughes Aircraft Company, Culver City, California

Springer Science+Business Media, LLC · 1972

This document has been approved for public release and sale ; its distribution is unlimited. Sponsored by : Air Force Materials Laboratory, Wright-Patterson Air Force Base, Ohio.

Library of Congress Catalog Card Number 76-147312
ISBN 978-1-4757-6003-3   ISBN 978-1-4757-6001-9 (eBook)
DOI 10.1007/978-1-4757-6001-9

©1972 Springer Science+Business Media New York
Originally published by **IFI/Plenum Data Corporation** in 1972.
Softcover reprint of the hardcover 1st edition 1972

FOREWORD

This report was prepared by Hughes Aircraft Company, Culver City, California under Contract No. F33615-70-C-1348. The work was administered under the direction of the Air Force Materials Laboratory, Air Force Systems Command, Wright-Patterson Air Force Base, Ohio, with Mr. B. Emrich, Project Engineer.

The Electronic Properties Information Center (EPIC) is a designated information Analysis Center of the Department of Defense, authorized to provide information to the entire DoD community. The purpose of the Center is to provide a highly competent source of information and data on the electronic, optical and magnetic properties of materials of value to the Department of Defense. Its major function is to evaluate, compile and publish the experimental data from the world's unclassified literature concerned with the properties of materials. All materials relevant to the field of electronics are within the scope of EPIC: insulators, semiconductors, metals, superconductors, ferrites, ferroelectrics, ferromagnetics, electroluminescents, thermionic emitters and optical materials. The Center's scope includes information on over 100 basic properties of materials; information generally regarded as being in the area of devices and/or circuitry is excluded.

Grateful acknowledgement is made for the review and comments of Dr. G.D. Cody of RCA Laboratories and Dr. B.W. Roberts of General Electric Co.

v

CONTENTS

ALPHABETICAL ALLOY INDEX

INTRODUCTION

The Electronic Properties Information Center has developed the Data Table as a compilation of the most reliable information available for the physical, crystallographic, mechanical, thermal, electronic, magnetic and optical properties of the many materials employed in the Solid State technology.

Data Tables originally served as an introduction to the graphic data compilations, published by the Electronic Properties Information Center as Data Sheets and issued by the National Technical Information Service, Springfield, Va., 22151. These include publications on Niobium (AD 608-398), Niobium Alloys and Compounds (AD 480 000), Niobium-Tin (AD 830 330, 838 460), Niobium-Zirconium (AD 804 473), Vanadium Silicide (AD 810 374) and Super-Conducting Thin Films (AD 704 554).

These Data Tables have proven valuable to the scientific and engineering community and increasing requests for this highly selective type of information prompted the Electronic Properties Information Center to bring information on the Niobium Alloys and Compounds up-to-date with this set of Data Tables.

Values for the range of parameters involved in superconductors are fully reported. These include primarily composition and material preparation, temperature and field strength. Our primary goal is not to compress the information available, but to select and present a rounded and fully representative view of the specific alloy.

This comprehensive review of each compound has been made possible by the extensive collection of documents in the EPIC files; to date over 47,000 technical journal articles and Government reports have been acquired by the Center. To compile this set of tables, over 1500 articles have been evaluated for relevant data.

As far as possible, the arrangement of data has been standardized and is presented in a consistent manner. The most fundamental property of a super-conductor appears to be its crystal structure so lattice parameters and symmetry are presented first. These are followed by the transition temperatures for the various compositions, the critical field, the critical current density, electronic specific heat and the magnetic susceptibility. The several theo-retical values such as coherence length, penetration depth, mean free path and Debye temperature, also the energy gap, as well as experimental data for the thermal conductivity and melting point. Various mechanical properties as hardness, ductility, shear strength and expansion coefficients are included for their importance to the technology of the material.

No data is given for non-superconductive compounds in an alloy system, (e.g. the niobium oxides) and certain non-superconducting alloys have been omitted. No data on high temperature phenomena is included.

The atomic % of the alloying material is given only if so stated in the original paper; the chemical formula may not be stoichiometric, but if the formula is given in the paper, then it is so reported in these tables.

The most highly valued aspect of this work is the fact that every individual data point is accompanied by a reference citation. This allows the reader to refer to the original research paper for additional information and in this way offers a representative bibliographic review of these alloys. Where two or more documents present the same data values, all are cited. The bibliography is arranged alphabetically by author, more than one document by the same author is distinguished by the letters A, B, C etc. A few citations are starred, these were added last and will be found in a supplementary bibliography at the very end.

The alloys have been arranged according to their position in the periodic table, but an alphabetical alloy index is given, immediately following the Table of Contents.

This set of Data Tables was begun by Mr. D. Grigsby, a member of the EPIC staff, but on his leaving, the work was completed by Mrs. M. Neuberger.

An extremely valuable source of additional information on niobium alloys and compounds is the Superconductive Materials Data Center, General Electric Research and Development Center, Schenectady, New York. References cited in Dr. B.W. Roberts, "Superconducting Materials and Some of Their Properties", National Bureau Of Standards Technical Note 482, May 1969, are used to provide data on niobium alloys and compounds as cited in this report. The Superconductive Materials Data Center is sponsored by the National Bureau of Standards, Office of Standard Reference Data.

It is only comparatively recently that major applications for superconductors have been sucessfully demonstrated. Of the various applications
presented in this chapter, only superconducting magnets and laboratory
instruments offer significant use of superconducting materials and principles.
An extensive literature review of superconducting devices covering the period
1959 to March 1967 was compiled by Goree and Edelsack (1).* The U.S. Department
of Commerce, National Bureau of Standards, Cryogenic Data Center, publishes
SUPERCONDUCTING DEVICES AND MATERIALS (2) which surveys the literature on a
quarterly basis. The Cryogenic Data Center has issued bibliographies on superconducting motors and generators (3), devices for measuring magnetic field
strength and direction (4), detectors (5), transformers (6), transmission lines
(7), amplifiers (8), and magnets (9). The references provided by these sources
and those included at the end of this chapter, will indicate to the reader of
this volume of the HANDBOOK OF ELECTRONIC MATERIALS, many of the benefits and
advantages inherent in superconductors.

Some of the exciting superconductor applications are in the field of
transportation and electrical power transmission. Sandia Laboratories have
successfully operated a test rocket sled at velocities up to 100,000 ft/sec
(3.05×10^4 m/s). The magnetic suspension system in this sled is, basically, a
superconducting coil attached to the moving vehicle with a shaped conductive
plate, embedded in the roadway, to act as a guide rail. Eddy currents are
produced in the plate when the magnet passes over it and these repel the
magnet to cause suspension. Superconducting magnets are useful in such systems
as the increased field that they provide, allows for a greater air gap to exist
between the rail and the test vehicle (10). Magnetic levitation has been
applied in wind tunnels (59) and is being considered by others for use in mass
transit systems. Speeds up to 300 mph are possible (11, 12).

A major use for superconductors in the aircraft industry, is for generators. Their advantage is in the significant weight reduction that may be
achieved. As an example of the financial aspect of such reductions, Hayden
(12) reports that the revenue-earning potential of the Boeing 707 or the BAC
VC 10 is about $240 per annum per pound of payload. For the supersonic
Concorde the figure should be about $600 per annum per pound of payload.
Using this figure he estimated that, if the Concorde's generator weight could
be halved, the potential revenue-earning capacity over the ten year life of the
aircraft would be increased by over $720,000. Such a weight reduction is possible
by using superconductive generators.

Electrical power transmission and other potentially important applications
are included in this chapter. There appears to be no technical reason why
superconductors should not become more widely used in commercial and aerospace
applications. Today's materials are adequate, the problems require only straight-
forward engineering techniques and the economics are reported to be very

---

* Numbers in parentheses refer to references listed on pages 14A-14C.

favorable (13). New developments are continually reported in both materials and technology. The first commercially available multifilament superconductor containing more than 400 highly stable superconducting niobium-titanium alloy filaments, each only 0.007 mm diameter was recently announced by Norton Co. (14). The filaments are embedded in a copper matrix for high thermal and electrical stability. The very small diameter enables the superconductor to achieve the variable high intensity magnetic fields used in nuclear research operations. It is capable of carrying pulsed or low frequency alternating current or direct current. The material called Supercon VSF makes possible varying magnetic fields up to 100 kGauss with negligible losses.

The near vacuum of outer space presents an ideal environment for superconducting devices in space vehicles. The double advantages of electrical power with space radiation shielding, indicate interesting possibilities for this application (15, 68).

It is very difficult to predict the future of superconducting materials applications in the light of present economic trends. Truly large scale use will require a significant break-through in the achievment of higher critical temperature superconductors, cryogenic cooling or a re-evaluation of those areas in which high magnetic fields and superconducting devices can be commercially useful (15). Until these problems have been solved from the economic view-point, they must be regarded as technical problems as well. Science, technology and economics have become so closely related that problems in one domain can barely exist without accompanying problems in the other regions (16).

COMPUTER DEVICES

Superconducting memories for computers have been under development for some time. The potential of superconductive Josephson tunneling technology for ultra-high performance memories and processors may surpass existing monolithic technology. Josephson tunneling circuits for memory and logic functions can be switched at subnanosecond delay, do not require standby power and dissipate extremely low energy, typically less than 10^{-13} Joules during fast switching operation. Indications are that 30 Megabit capacity, less than 1 Watt refrigeration (at 3.6°K) and cycle times of 40 ns to 15 ns can be achieved. One possibility is reported by Matisoo (21), using lead, tin and tin oxide. Niobium is used as the ground plane insulation (18, 19, 20, 21).

One possibility for research lies in the application of Josephson junctions in the control and movement of single quantized vortex lines as computer elements. Calculations by Anderson indicate great advantages in size and speed (22). Although research and development work on superconductive memories is still continuing in this country, Europe and Japan, the difficulties in fabrication are formidable, and the feasibility and utility of these computers is still uncertain. Developments in the programming, economics and technology of current computers may, in effect, design around the potential advantages of cryogenic memories before the development can be the basis for commercial use (17).

4

FLUX PUMPS

The capability of superconducting inductors to store energy in magnetic flux has led to various transfer devices which shift flux from one inductor to another. These transfer devices, flux pumps, are capable of generating large direct currents from electrical or mechanical inputs (23). Flux pumps have been developed for providing large currents of 1,000 to 10,000 Amps for superconducting magnets. The use of flux pumps is a factor in minimizing the thermal losses associated with running heavy electrical leads from room temperature into the cryogenic environment (17, 24).

GYROSCOPES

Gyroscopes based on the use of superconductors have been under development for a number of years and are now undergoing flight tests (17, 25, 61). The two principal types of superconductor gyroscopes which would be of value to navigation are the superconductive-rotor gyro and the cryogenic nuclear gyro. The inherent advantage of the superconductive device is that it does not require feedback circuits for stability as does the electrostatic gyroscope.

The superconductive-rotor gyroscope is designed to provide random drift rates as low as 0.0001°/hour. The superconducting rotor is supported in a vacuum by a magnetic field. As a steady current is established in the rotor, it spins at speeds greater than 30,000 rpm without attenuation. Hollow niobium spheres or rotors coated with thin superconductive films have been used. Problems of heat transfer and sealing remain to be resolved (12).

The cryogenic nuclear gyroscope is based upon the intrinsic angular momentum of particular atomic nuclei as a substitute for the conventional gyro rotor. The device requires shielding from undesirable external magnetic fields which interfere with the atoms. Superconductive materials are being evaluated to provide the shielding for this device.

Although potential advantages are foreseen, they have not been fully realized; in the meantime other types of gyros are rapidly approaching the performance predicted for the superconducting gyros.

INSTRUMENTS

Superconductors have long been used in low temperature environments. Niobium wires are used in electrical connections where thermal isolation of a sample is desired. In such applications the wires are superconducting with a high critical field throughout the entire liquid helium temperature range and combine low thermal transport with perfect electrical conductivity (18).

Superconducting ultra-sensitive amplifiers possessing relatively low-frequency response and very low input impedance are well suited for studying low temperature phenomena associated with low impedance sources (26, 64, 65). An extremely sensitive detector of this type is the superconducting galvanometer developed by Pippard and Pullan. By using single-turn deflection coils for low inductance and a low controlling field of about 0.01 Gauss for high sensitivity, this instrument was able to detect 10^{-5} Amps corresponding to an emf of 10^{-12} Volts, with a time constant of 15 sec. A comparison of low-signal amplifiers is given in Table I:

TABLE I

Comparison of Representative Superconducting Low-Signal Amplifiers

Amplifier Type	Minimum detectable voltage, V	Bandwidth Hz	Minimum detectable current, A	Quality Factor $V_{min}B^{-1.5}$ V/cycles$^{1.5}$	Ref.
Galvanometer	10^{-12}	10^{-2}		10^{-9}	27
Wire Cryotron Modulator	10^{-11}-10^{-12}	0.16		10^{-10}-2×10^{-11}	28
Cascaded Film Cryotron Bridge	5×10^{-11}	45×10^3	7×10^{-8}	5×10^{-18}	29
Josephson Junction	5×10^{-14}	0.8	10^{-6}	7×10^{-15}	30

Clark has constructed devices utilizing transformers with which about 10^{-15} V (with Josephson noise of about 5×10^{-16} V) have been detected (31).

Superconducting magnetometers with a sensitivity of 10^{-8} Gauss have been developed (32), and are used for low level measurements in a low-temperature low-noise environment. By using superconducting transformers and appropriate shielding, Beasly and Webb (57) have developed a magnetometer which may be used in a high-field background. The device can resolve changes of 10^{-7} Gauss with one second time constant in a field of 2500 Gauss, a field resolution of less than 1 in 10^{10}.

Cohen et al. (33) have used a superconducting point-contact magnetometer to take magneto-cardiograms inside a shielded room. These magneto-cardiograms approach good medical electro-cardiograms in clarity and are an order of magnitude improvement in sensitivity over previous magnetic heart detectors. The results suggest new medical uses for this superconducting magnetometer.

Because of their strong variation with temperature, superconductors make excellent infrared detectors. A current-carrying niobium nitride ribbon held in the mid-point of its transition was developed so that incident radiation produced a voltage change across the detector. This device could detect a light flux of 5×10^{-10} Watt or an individual flash of light having an energy of 2×10^{-13} Joule (34).

Since the discovery of the Josephson effect, increased sensitivity has become possible. A niobium powder, Josephson junction has been developed in which the oxidized niobium acts as the insulating part of the junction. The device has a sensitivity of 10^{-14} Watt/Hz$^{0.5}$ and will be used in a radio tele-scope. The Josephson junction is an optimal astronomical detector for wave-lengths between 100 microns and 3 mm and possibly longer wavelengths; in this range the device offers high sensitivity, subnanosecond response times and insensitivity to near infrared and visible radiation (35). Josephson junctions are also used as microwave mixers (66) and in noise thermometry (67).

Superconducting high field coils have been used successfully in electron microscopes. Hitachi Ltd. has developed such a microscope that has a resolution of about 30 Å when the device is operating at an accelerated voltage of 400 kV and the peak value of the axial magnetic field of the lens is 45 kGauss (36).

Scientists at the National Bureau of Standards, Institute for Basic Research (37) have accurately measured the voltage obtained from a Josephson junction irradiated by microwaves of a know frequency. Accurate frequency-to-voltage conversions exhibited by the Josephson junction hold very significant promise for surveillance measurements of standard celss, possible applications as standard voltages and more accurate determinations of fundamental constants such as electron charge, Planck's constant and electron rest mass. Other standards laboratories will be able to make their own measurements and eliminate the need for frequent comparison with the NBS standard cells.

LINEAR ACCELERATORS

A departure from conventional linear accelerator design is utilized at the 3 km long Stanford Linear Accelerator and at Karlsruhe, Germany to generate protons of 1 GeV energy and provides another application for superconductivity. The technology used here reduces energy losses by the application of superconductivity to the walls of the resonant accelerator sections which carry the radio-frequency currents generated by the klystrons. The radio-frequency power requirement is a major factor in the construction budget of an accelerator. The economic balance can be shifted radically, if the resonators, which form the linear accelerator are coated with superconducting material. In this case, practically all the radio-frequency power is delivered to the beam and the rf equipment no longer plays an important part in the budget of the accelerator. The cost of the cooling system now takes over as the main component of the total investment.

The most promising materials for coating the resonators are lead and niobium; the former is reported easier to use (38). Experimental results on the decay of oscillations in pulsed superconducting cavities, showed that there is a contribution to surface resistance which depends on the surface preparation. Until the present, low-resistance lead surfaces have not been possible but niobium, under study at Stanford, shows surface resistance near to theoretical values under certain circumstances.

Wilson and co-workers (39) at the Stanford Linear Accelerator Center have presented a survey of research and development directed to converting the present machine to a two-mile 100 GeV superconducting Linac. Technological problems which hamper the conversion are being handled by two different approaches: development of a better understanding of superconducting materials and construction of a short superconducting accelerator. The latter approach utilizes a 15-cavity niobium traveling-wave resonant ring structure operating in the $2\pi/3$ mode. The frequency, (2856 MHz) and the accelerating field (33 MV/m) are the same for a short accelerator as is envisaged for the two-mile machine.

POWER GENERATION

The advantage of using superconductors in power generating machinery is the reduction in size and weight made possible by decreasing or eliminating the use of iron or magnetic flux paths. There are, therefore, obvious applications for airborne or space power plant generators, mobile field generators, ship propulsion systems and large electric motors and generators for conventional industrial applications (13). Power requirements for high level continuous power, high power pulses and high level continuous power with pulsing capabilities may be achieved by utilizing superconductor materials.

Continuous high level power in the MegaWatt and higher range can be provided by magnetohydrodynamic (MHD) power generation systems. Superconductor materials are extremely important to MHD propulsion systems which employ magnetic means to confine and direct ion streams. Hot gas from the system combustion chamber flows through a nozzle at high velocity and in so doing, cuts the transverse magnetic lines of force, thereby generating a current. This current is then picked up by the electrodes and flows to a power distribution system. The power generated is proportional to the square of the magnetic field. The advantages of MHD include the ability to start and reach full power in several seconds (40,41).

Fushimi (36) in reporting on superconducting magnet applications in Japan has described the successful operation of MHD power generator with a superconducting magnet. The saddle-shaped magnet was wound with a superconducting strip of AVCO SG-500 and produced a central field of 23 kGauss. The maximum output was 25 W (25 V, 1 A).

Superconducting solenoids are also used for energy storage devices; the energy stored is proportional to the square of the field. Such induction energy storage devices may be used where high energy pulses of short duration and high cycling rates are required. Materials problems have been found with such devices in the eddy current losses in the metallic walls of the magnet's cooling container.

The third power generating application is in rotating machinery. The principles of such a device are the same as those of a conventional generator. The generated voltage is proportional to the armature length, the relative velocity and the magnetic field strength. There are two possible configurations; one in which the magnet rotates and the armature is stationary and the other where the magnet is stationary and the armature rotates. The rotating field configuration has the advantage that the mechanic stress is reduced, since the smallest member rotates and the electrical connections are stationary. The major problem of the configuration is that the moving member must be cooled. The seal for such a transfer may be a serious problem.

The most advanced development to date uses the winding simplicity of an axial solenoidal field, instead of a radial multipole field. Such a homopolar machine has been successfully operated at 50 kW. A three-250 hp, 200 rpm motor has been designed and is now being successfully operated by the International Research and Development Co., Ltd. at Fawley Power Station in England (42).

8

A comparison of costs between a normal and a superconducting motor was presented by International Research and Development Co. Ltd. (43) as follows:

	Normal Motor	Superconducting Motor
Horsepower	8000	8000
Speed	50 rpm	50 rpm
Weight	290 tons	40 tons
Cost*	$420,000	$312,000
Overall Efficiency	94%	97%

* Based on estimates of 1970 costs.

Woodson and Thullen of the Massachusetts Institute of Technology (44) have developed a prototype two-pole, 3,600 rpm design rated at 80 kVA, 450 A/phase alternating current generator. The active part of the rotor is enclosed in a vacuum to reduce heat losses and the superconducting windings of niobium-titanium are maintained by liquid helium at 4°K. Their trial designs for two- and four-pole machines with both iron and copper magnetic shields, indicated that the initial specific costs for the superconducting section of the machine, including the refrigeration system, would be from 0.30 to 0.10 $/kVA over the 1,000 to 10,000 MVA range.

POWER TRANSMISSION

One of the most important potential applications for superconductors is the transmission of electrical power. Serious attention has been given in the U.S. and Europe to the technical feasibility, economics and marketability of cryogenic power transmission lines with superconductors or ordinary conductors. Garwin and Matisoo (45) presented a detailed analysis of the advantages and problems associated with such a system. They proposed a superconductor cable made up of twin lines carrying a direct current of 100 kAmps. with a voltage of ±20 kV between the conductors. The two conductors are enclosed in a common tube of 12 cm diameter, whose temperature is maintained at 4.2°K. The helium filled tube is thermally isolated by a vacuum zone against its surroundings, which are cooled to 77°K by liquid nitrogen. The whole system fits into a vacuum tight metal tube; this is embedded in a concrete channel, to act as a support. The metal trough has a cover which may be lifted in small sections for inspection and repair. The vacuum line for such a system as discussed by Paul (46) is subdivided into sections and the proposed refrigeration system is shown. Moisson and Leroux (62) in France, report on a program for underground cryogenic aluminum cables. Another recent discussion is given by Long and Notaro (63). The problems of heat loss are extremely important. If the cost of helium refrigeration for superconductor power transmission lines could be halved, the only problem standing in the way of commercial utilization would be vacuum maintenance (47, 48). Development of microwave lines using circular waveguides in the TE_{01} mode at frequencies from 3 to 10 GHz, pose competition for power transmission by superconductor systems (46).

9

Capital costs rather than operating costs are dominant in the economics of power transmission (49). Figure 1 shows a cost comparison of the various power transmission systems of the future. Although capital costs are of prime importance, it is also necessary to minimize losses during transmission. The Edison Electric Institute has estimated that a 16 km, 138 kV, 1690 MVA superconducting line would cost about $550/MVA-km, including capitalized operating costs (50). Linde Division, Union Carbide Cor. (51) has announced a 12-year development program to place a superconducting cable in operation by 1981. Research reported by Dr. Long shows that an electro-deposited niobium sample with a 1 cm diameter and 0.05 mm thick, can carry 1,700 Amps. At this current density of more than 100,000 Amps/cm^2, the alternating current losses are only about 4.5 mW/m. The largest underground conventional cables currently in use have a current rating of less than 1,000 Amps. At a current density less than 100 Amps/cm^2 they show a loss of about 30 W/m. This is 6,000 times that of the superconducting cable system.

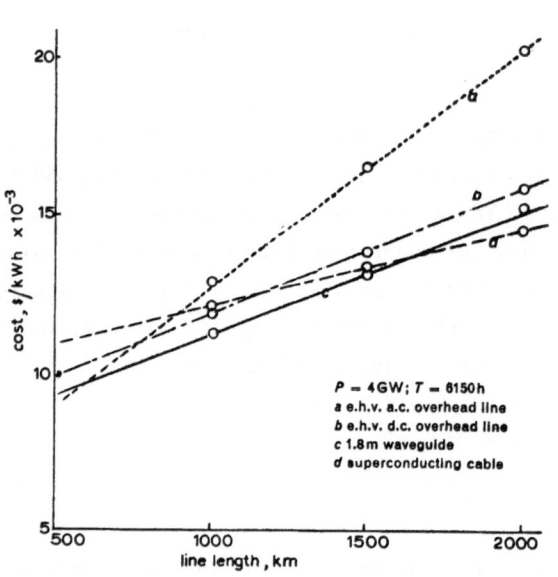

P = 4GW; T = 8150h
a e.h.v. a.c. overhead line
b e.h.v. d.c. overhead line
c 1.8m waveguide
d superconducting cable

Cost estimates of delivered power for different transmission systems (46)

SUPERCONDUCTING MAGNETS

The most important application of superconductivity today is the generation of large volume and high dc magnetic fields by using niobium-zirconium, niobium-titanium alloys or niobium stannide. These superconducting magnets are the key component in the various applications discussed below (58). Large superconducting magnets are especially attractive for high energy and plasma physics because of their relatively low power requirements and space occupied (52). The capital and operating costs of superconducting magnets are much less than those for normal magnets. Immediate applications include beam-bending and focussing magnets (43). Superconducting magnets are already in operation to provide stable confinement for hydrogen plasma for controlled thermonuclear fusion (Norton Co., 60).

Superconducting magnets have three limiting features:
1. A cryogenic environment is required for start-up and operation,
2. If improperly operated, they may revert to the resistive state,
3. Field strength can not be easily varied; the heat produced boils off the coolant almost immediately and may damage the magnet. The field strength of conventional electromagnets can be varied more easily and quickly than that of superconductive magnets employing persistent current operation.
These factors bar some potential industrial applications of superconductivity (41).

A superconducting quadropole focussing magnet constructed by the Oxford Instrument Co., Ltd. successfully completed its final trials early in 1970. A field gradient of 5.5 kGauss/cm^2 in a space of 10 cm diameter by 70 cm is produced. This is achieved with a current density through the windings of 10,000 Amps/cm^2 and a conductor current of 820 Amps. The peak field at the winding is 50 kGauss. The superconducting material used is a multi-core, niobium-titanium/copper composite, manufactured by IMI Ltd. (53).

For synchrotron magnets, emphasis centers on composite conductors of many fine filaments of niobium-titanium in a copper or cupronickel matrix, with the strands suitably transposed; for a typical pulse frequency of 1 Hz, the superconductor strands must be thinner than 10 microns at which the losses with a peak excitation of 6 Tesla may be typically less than 0.06 J/cm^3 (54). One of the remaining problems is devising an acceptable coil construction. Currently, full impregnation of the winding with epoxy resin is preferred. The most interesting synchrotron application is the 500 cm diameter superconducting ring at the Culham Laboratory Levitron. The superconducting ring in a persistent current mode will float in the field produced by the outer coil system.

Superconducting magnets used in Argonne National Laboratory, 12 ft. diameter bubble chamber have been in operation for over a year. These magnets use fully stabilized conductors which are already obsolete (55).

Superconducting electromagnets have been used to provide a stable field of several kGauss for masers at the Massachusetts Institute of Technology, Bell Telephone Laboratories and Westinghouse Electric Corp. A 70 GHz travelling-wave maser with field coils wound of superconducting wire, has been developed at Westinghouse. A 5 kGauss field is produced by a niobium-zirconium super-conducting magnet with a deviation of less than 1 Gauss, perpendicular to the 1.5 inch length of the travelling-wave maser element. Energy storage devices capable of delivering up to several hundred Joules, have been built to pump lasers. Material costs make energy storage by superconductors uneconomic at energies of less than 10^7 Joules (41).

A 100 kOe (8x10^6 Amps/m) superconducting coil magnet has been operated at the Kamerlingh Onnes Laboratory, The Netherlands (56). It is energized by means of a 100 Amp. 10 V power supply outside of the cryostat. The coil comprises three concentric sections, each 140 mm long. The outer section is wound from niobium-titanium wire. The middle and inner sections are wound from niobium stannide ribbon. The entire magnet cost $12,600. If niobium stannide had been used for the outer section as well, the magnet would have been lighter and required less helium coolant, but would have cost $22,000.

REFERENCES

1. GOREE, W.S. and E.A. EDELSACK. Superconducting Devices - A Literature Survey. STANFORD RESEARCH INSTITUTE, Menlo Park, California. Mar. 1967. 68 p. Available NTIS* as AD 651 376.

2. Superconducting Devices and Materials - A Literature Survey Issued Quarterly. CRYOGENIC DATA CENTER, NATIONAL BUREAU OF STANDARDS, Boulder, Colorado.

3. Superconducting Motors and Generators. CRYOGENIC DATA CENTER, NATIONAL BUREAU OF STANDARDS, Boulder, Colorado. Bibliography B-638. Sept. 1970. 23 p.

4. Superconducting Devices for Measuring Magnetic Field Strength and Direction. NATIONAL BUREAU OF STANDARDS, CRYOGENIC DATA CENTER, Boulder, Colorado. Bibliography B-639, Sept. 1970. 13 p.

5. Superconducting Detectors. NATIONAL BUREAU OF STANDARDS, CRYOGENIC DATA CENTER, Boulder, Colorado. Bibliography B-640. Sept. 1970. 21 p.

6. Superconducting Transformers. NATIONAL BUREAU OF STANDARDS, CRYOGENIC DATA CENTER, Boulder, Colorado. Bibliography B-641. Sept. 1970. 13 p.

7. Superconducting Transmission Lines. NATIONAL BUREAU OF STANDARDS, CRYOGENIC DATA CENTER, Boulder, Colorado. Bibliography B-642. Sept. 1970. 22 p.

8. Superconducting Amplifiers. NATIONAL BUREAU OF STANDARDS, CRYOGENIC DATA CENTER, Boulder, Colorado. Bibliography B-643. Sept. 1970. 12 p.

9. Superconducting Magnets. NATIONAL BUREAU OF STANDARDS, CRYOGENIC DATA CENTER, Boulder, Colorado. Bibliography B-644. Sept. 1970. 152 p.

10. Magnetically Suspended Rocket Sled. CRYOGENICS, v. 10, no. 2, Apr. 1970. p. 171.

11. Magnetic Levitation. INDUSTRY WEEK, v. 167, no. 12, Sept. 21, 1970. p. 24.

12. HAYDEN, J.T. Superconductivity and Possible Applications in Transportation. ELECTRONICS AND POWER, v. 15, Aug. 1969. p. 281-283.

13. TAYLOR, C.E. Future Prospects for Applications of Superconductivity. CRYOGENICS, 69 CONFERENCE, June 1969. 16 p.

14. First AC or DC Superconductor. MATERIALS ENGINEERING, v. 72, Nov. 1970. p. 27.

15. ANASHKIN, O.P. et al. Test of a Superconducting Magnetic System with a Field H ∿ 20 kOe on Board a Satellite. KOSMICHESKIE ISSLEDOVANIIA, v. 7, Sept.-Oct. 1969. p. 786-793.

16. SCHWARTZ, B.B. Pessimists and Optimists Discuss Superconductivity at Stanford. PHYSICS TODAY, v. 23, no. 4, Apr. 1970. p. 79-83.

17. SCHMITT, R.W. and W.A. MORRISON. Economic Aspects of Superconductivity. In: SUPERCONDUCTIVITY IN SCIENCE AND TECHNOLOGY, Edited by M.H. Cohen. UNIVERSITY OF CHICAGO PRESS, Chicago, Illinois. 1968. p. 127-155.

18. LYNTON, E.A. Superconductivity. METHUEN & CO., LTD, London, England. 1969. 219 p.

19. COHEN, M.H. (Editor). Superconductivity in Science and Technology. UNIVERSITY OF CHICAGO PRESS, Chicago, Illinois. 1968. 163 p.

20. ANACKER, W. Potential of Superconductive Josephson Tunneling Technology for Ultrahigh Performance Memories and Processors. IEEE TRANS. ON MAGNETICS, v. MAG-5, no. 4, Dec. 1969. p. 968-975.

21. MATISOO, J. Josephson-Type Superconductive Tunnel Junctions and Applications. IEEE TRANS. ON MAGNETICS, v. MAG-5, no. 4, Dec. 1969. p. 848-873.

22. ANDERSON, P.W. How Josephson Discovered His Effect. PHYS. TODAY, v. 23, no. 11, Nov. 1970. p. 23-29.

23. HULM, J.K. Superconductors in Technology. In: SUPERCONDUCTORS IN SCIENCE AND TECHNOLOGY, Edited by M.H. Cohen, UNIVERSITY OF CHICAGO PRESS, Chicago, Illinois. 1968. p. 103-126.

24. HULM, J.K. and D.W. DEIS. Applications of Superconductivity. ELECTRO-TECHNOLOGY, v. 84, no. 1, July 1969. p. 57-66.

25. RAIBOV, A.B. Calculation of the Force Characteristics of a Single-Coil Suspension of the Rotor of a Cryogenic Gyroscope. PRIBOROSTROENIE, v. 13, no. 2, 1970. p. 86-90.

* NTIS, National Technical Information Service, Springfield, Va. 22151.

26. NEWHOUSE, V.L. Superconducting Devices. In: SUPERCONDUCTIVITY, edited by R.D. Parks. Volume 2, MARCEL DEKKER, INC., New York, New York, 1969. p. 1283-1342.

27. PIPPARD, A.B. and G.T. PULLAN. A Superconducting Galvanometer. CAMBRIDGE PHILOSOPHICAL SOC., PROC., v. 48, 1952. p. 188-194.

28. TEMPLETON, I.M. A Superconducting Reversing Switch. J. OF SCI. INSTRUMENTS, v. 32, May 1955. p. 172-173.

29. NEWHOUSE, V.L. and H.H. EDWARDS. An Ultrasensitive Linear Cryotron Amplifier. IEEE PROC., v. 52, Oct. 1964. p. 1191-1206.

30. CLARKE, J. A Superconducting Galvanometer Employing Josephson Tunnelling. PHIL. MAG., v. 13, no. 121, Jan. 1966. p. 115-128.

31. CLARKE, J. The Measurement of Small Voltages Using a Quantum Interference Device. In: SYMPOSIUM ON THE PHYSICS OF SUPERCONDUCTING DEVICES, UNIV. OF VIRGINIA, Charlottesville, Virginia, Apr. 28-29, 1967. p. D-1 to D-12. Available NTIS* as AD 661 848.

32. FORGACS, R.L. and A. WARNICK. Digital-Analog Magnetometer Utilizing Superconducting Sensor. REV. OF SCIENTIFIC INSTRUMENTS, v. 38, no. 2, Feb. 1967. p. 214-220.

33. COHEN, D. et al. Magnetocardiograms Taken Inside a Shielded Room with a Superconducting Point-Contact Magnetometer. APPLIED PHYS. LETTERS, v. 16, no. 7, Apr. 1, 1970. p. 278-280.

34. ANDREWS, D.H. et al. A Fast Superconducting Bolometer. OPTICAL SOC. OF AMERICA, J., v. 36, no. 9, Sept. 1946. p. 518-524.

35. Josephson Detectors Make Astronomical Observations. PHYSICS TODAY, v. 23, no. 4, Apr. 1970. p. 55-56.

36. FUSHIMI, K. Applications of Superconducting Magnets in Japan. CRYOGENICS, v. 10, Apr. 1970. p. 116-118.

37. Josephson Voltage Seen as Potential Standard. INDUSTRIAL RESEARCH, v. 12, no. 13, Dec. 1970. p. V1.

38. CITRON, A. The Present Status of Superconducting Linear Accelerators. CRYOGENICS, ICEC 3, Supplement, June 1970. p. 12.

39. WILSON, P.B. et al. Superconducting Accelerator Research and Development at SLAC, in Particle Accelerators, v. 1, 1970. GORDON & BREACH, SCIENCE PUBLISHERS LTD., Glasgow, Scotland. p. 223-238.

40. KUHL, G.E. Superconductors for Air Force Applications. Paper presented at the 1970 AIR FORCE MATERIALS SYMPOSIUM, May 18-22, 1970, Miami Beach, Florida. 9 p.

41. FOX, D.K. Putting Superconductors to Work. ELECTRONICS, v. 39, no. 3, Feb. 7, 1966. p. 95-101.

42. MacNAB, R. and A.D. APPLETON. The Superconducting Homopolar Motor. Paper presented at the Third INTERNAT. CRYOGENIC ENGINEERING CONF., May 25-27, 1970. West Berlin. Brief Summary, CRYOGENICS, v. 10, no. 4, Aug. 1970. p. 343.

43. BEATSON, C. Developing Future Commercial Uses of Superconductivity. THE ENGINEER, v. 229, no. 5934, Oct. 16, 1969. p. 35-39.

44. Superconducting Winding for Large Synchronous Generator. ELECTRICAL REVIEW, v. 187, no. 10, Sept. 4, 1970. p. 343-344.

45. GARWIN, R.L. and J. MATISOO. Superconducting Lines for the Transmission of Large Amounts of Power Over Great Distances. In: SUPERCONDUCTORS IN SCIENCE AND TECHNOLOGY, Edited by M.H. Cohen, UNIV. OF CHICAO PRESS, Chicago, Illinois. 1968. p. 77-91.

46. PAUL, H. Power Transmission of the Future-Microwaves on Superconductors. ELECTRONICS AND POWER, v. 16, May 1970. p. 171-174.

47. Industrial Cryogenics: Hot Competition, Huge Potentials. CRYOGENICS & INDUSTRIAL GASES, v. 5, no. 1, Jan. 1970. p. 31-32.

48. FOX, M. The Production and Maintenance of Vacuum in Superconducting Power Cables. VACUUM, v. 20, no. 3, Mar. 1970. p. 97-107.

49. Superconducting Power Transmission. PHYSICS TODAY, v. 23, no. 12, Dec. 1970. p. 42-43.

50. EEI Research Project on Superconducting Cable Systems. EDISON ELECTRIC INSTITUTE, BULLETIN, May 1969. p. 158.

51. Superconducting Power Cables in 12 Years. INDUSTRIAL RESEARCH, v. 11, no. 8, Aug. 1969. p. 19-20.

52. FISHLOCK, D. A Guide to Superconductivity. MacDONALD & CO. LTD, London, England, 1969. 160 p.

53. Magnet Completes Test. CRYOGENICS, v. 10, no. 2, Apr. 1970. p. 171.

54. BRONCA, G. Progress in ac Superconducting Magnets. Paper presented at the Third INTERNAT. CONF. ON MAGNETIC TECHNOLOGY, May 19-22, 1970, Hamburg, Germany. Paper Discussion in CRYOGENICS, v. 10, no. 4, Aug. 1970. p. 337.

55. MULHALL, B.E. and A.D. APPLETON. Third International Conference on Magnet Technology. CRYOGENICS, v. 10, no. 4, Aug. 1970. p. 336-337.

56. METSELAAR, J.W. A 100 kOe Superconducting Coil Magnet. CRYOGENICS, v. 10, no. 3, June 1970. p. 220-223.

57. BEASLEY, M.R. and W.W. WEBB. Operation of Superconducting Interference Devices in Appreciable Magnetic Fields. SYMPOSIUM ON THE PHYSICS OF SUPERCONDUCTING DEVICES, UNIV. OF VIRGINIA, Charlottesville, Virginia, Apr. 28-29, 1967. p. V-1 to V-8. Available NTIS as AD 661 848.

58. STEKLY, Z.J.J. State of the Art of Superconducting Magnets. J. OF APPLIED PHYS., v. 42, no. 1, Jan. 1971. p. 65-72.

59. ZAPATA, R.N. et al. University of Virginia Superconducting Wind-Tunnel Balance. J. OF APPLIED PHYS., v. 42, no. 1, Jan. 1971. p. 3.

60. COLES, W.D. et al. Superconducting Magnet Research. CRYOGENICS AND INDUSTRIAL GASES, v. 5, no. 9, Nov.-Dec. 1970. p. 15-20.

61. MILLER, B. Navaid Uses Electrostatic Gyros. AVIATION WEEK & SPACE TECHNOLOGY, v. 94, no. 12, Mar. 22, 1971. p. 49.

62. MOISSON, F. and J.M. LEROUX. Development of a Superconducting Cable for Transmission of High Electric Power. J. OF APPLIED PHYS., v. 42, no. 1, Jan. 1971. p. 154.

63. LONG, H.M. and J. NOTARO. Design Features of ac Superconducting Cables. J. OF APPLIED PHYS., v. 42, no. 1, Jan. 1971. p. 155-162.

64. LUKENS, J.E. et al. Versatile Superconducting Femtovolt Amplifier and Multimeter. J. OF APPLIED PHYS., v. 42, no. 1, Jan. 1971. p. 27.

65. BERTIN, C.L. and K. ROSE. Comparison of Superconducting and Semiconducting Bolometers. J. OF APPLIED PHYS., v. 42, no. 1, Jan. 1971. p. 163-166.

66. DiNARDO, A.J. and E. SARD. Superconducting Microwave Mixes Utilizing Josephson Junctions. J. OF APPLIED PHYS., v. 42, no. 1, Jan. 1971. p. 105.

67. KAMPER, R.A. and J.E. ZIMMERMAN. Noise Thermometry with the Josephson Effect. J. OF APPLIED PHYS., v. 42, no. 1, Jan. 1971. p. 132-133.

68. TESDALL, D.W. Application of Superconducting Devices in Space. IEEE TRANS. ON MAGNETICS, v. MAG-5, Sept. 1969. p. 433-434.

NIOBIUM HYDROGEN

PROPERTIES	SYMBOL	VALUE	UNIT	AT.% H	CRYSTAL DATA	NOTES	REFERENCES
Lattice Parameters	a_o	3.304	Å	0	bcc		Edwards et al.
		3.311		5.06	α-bcc	cooled from 800°C	Horn & Ziegler
		3.327		9.89	α-bcc		
		3.43		50.0	β-bcc		Samsonov & Anmonova
		4.55		67.0	β-bcc		
Transition Temperature	T_c	8.98	°K	0	bcc		Horn & Ziegler
		7.83		5.06	α-bcc	cooled from 800°C	
		7.38		9.89	α-bcc		
		7.28		32.76	α-bcc		
		12.7		50.0	β-bcc		Aschermann

NIOBIUM GOLD

PROPERTIES	SYMBOL	VALUE	UNIT	AT.% Au	CRYSTAL DATA	NOTES	REFERENCES
Lattice Parameters	a_o	5.2027	Å	Nb_3Au	A-15	arc-melted, annealed in vacuo, 1 week-1400°C	Zegler, Wood & Matthias
		3.29		Nb_3Au	A-2		Bucher et al. E
Transition Temperature	T_c	8	°K	15	A-15	arc-melted	Bucher et al. E
		8.4		20			
		11.1		25			
		8.2		30			
		11.5		Nb_3Au	A-15	arc-melted	Zegler
		6.0		5	A-2	arc-melted, quick-quenched	Bucher et al. E
		4.0		10	A-2		
		1.2		25	A-2		
Magnetic Susceptibility	$\chi_{gr.at.}$	1.65	10^{-6}emu	Nb_3Au		extrapolated to 0°K	Bernasson et al.
		1.614				at 77°K	Bucher et al. E
		1.487				at 300°K	

NIOBIUM MAGNESIUM

PROPERTIES	SYMBOL	VALUE	UNIT	AT.% Mg	CRYSTAL DATA	NOTES	REFERENCES
Transition Temperature	T_c	5.6	°K	$NbMg_2$			Alexseeskii & Mikhailov

NIOBIUM BORON

PROPERTIES	SYMBOL	VALUE	UNIT	AT.% B	CRYSTAL DATA	NOTES	REFERENCES
Lattice Parameters	a_o	4.210	Å	10	β		Andersson & Kiessling
	a_o	3.297		NbB	orthorhombic	pressed powder, heated 24 hours, 1300°C	Tyan et al.
	b_o	8.716				annealed, 24 hours 1650°C	
	c_o	3.164					
	a_o	3.305		Nb_3B_4			Darnell & Yntema
	b_o	14.08					
	c_o	3.137					
	a_o	3.085		NbB_2	hexagonal		Brewer et al.
	c_o	3.311					
Transition Temperature	T_c	8.25	°K	NbB	orthorhombic		Matthias & Hulm
		6.94		NbB	orthorhombic	electron-beam melted and zone-refined	Gaule et al.
		6.4		55		sintered in argon, 1700-1750°C	Gaule et al.

PROPERTIES	SYMBOL	VALUE	UNIT	AT.% B	CRYSTAL DATA	NOTES	TEMP.(°K)	REFERENCES
Transition Temperature	T_c	6.0		NbB				Hulm & Matthias
		1.27		Nb_3B_4				Matthias & Hulm
		1.27		NbB_2	hexagonal			
Critical Field	H_c	4.45	kGauss	NbB	β-MoB type	electron-beam melted, zone-refined	4.2	Gaule et al.
		8.00		55		sintered in argon at 1700-1750°C	4.2	Gaule et al.
		4.8		59.3				
Electronic Specific heat	γ	1.55	mJ/mole °K^2	NbB			0	Tyan et al.
		4.0		NbB_2		arc-melted	1.0	Ukei & Kanda
		0.5					0.7	
Electrical Resistivity	ρ	9.72	$\mu\Omega$-cm	NbB	orthorhombic	electron-beam melted, zone-refined, $\rho T_c/\rho_{300} = 0.0261$		Gaule et al.
		8.120		55		sintered in argon at 1700-1750°C, $\rho T_c/\rho_{300} = 0.0345$		Gaule et al.
		14.76		59.3		sintered in argon at 1700-1750°C, $\rho T_c/\rho_{300} = 0.0366$		Gaule et al.
		32.0		NbB_2			300	Kendall & McClelland, Lvov et al.
Debye Temperature		566	°K	NbB			0	Tyan et al.
		720		NbB_2			300	Samsonov & Grebenkina
Thermal EMF		-1.4	μV/°C	NbB_2			300	Lvov et al.
		-3.7		NbB_2		arc-melted	300	Davisson et al.
		-1.2				annealed		
Thermal Conductivity		0.17	W/cm °K	NbB_2				Gambino
		0.167					300	Kendall & McClelland
		0.197-0.259					493	

PROPERTIES	SYMBOL	VALUE	UNIT	AT.% Al	CRYSTAL DATA	NOTES	TEMP.(°K)	REFERENCES
Lattice Parameters	a_o	5.192	Å	17-20	$Nb_3Al+\alpha$-Nb			Kunz & Saur A
		5.180		24.2	Nb_3Al			
		5.187		25	A-15	pressed powder fired in Helium		Wood et al.
		5.189		25	Nb_3Al+Nb_2Al	sintered 1000°C crushed and pressed 20°C annealed 6 hours 1500°C		Bachner & Gatos
	a_o	5.167		33	Nb_2Al,			Gladyshevskii et al.
	c_o	9.957			σ-tetragonal			
	a_o	5.438		75	$NbAl_3$,			Hansen
	c_o	8.601			tetragonal			
Transition Temperature	T_c	18.8	°K		Nb_3Al	high purity pressed powder, solid state reacted and annealed at 700°C, a_o=5.182 Å		Willens et al.

PROPERTIES	SYMBOL	VALUE	UNIT	AT.% Al	CRYSTAL DATA	NOTES	REFERENCES
Transition Temperature	T_c	18.3	°K	Nb_3Al		fused, annealed 1 hour 900°C	Meyer A
		17.3		Nb_3Al		arc-melted	Otto
		18.4				annealed 10 hours 1150°C	
		16.4		Nb_3Al		sintered 1000°C crushed and pressed, annealed 6 hours 1500°C	Bachner & Gatos
		17.20		Nb_3Al		arc-melted, induction meas.	Kunz & Saur B
		18.22				annealed 1 hour 790°C	
		17.30				arc-melted, resistivity meas.	
		18.28				annealed, 1 hour 790°C	
		17.07		27.3		arc-melted, induction meas.	
		18.07				annealed, 1 hour 800°C	
		17.7		Nb_3Al		arc-cast	Swartz et al.
		17.48				irradiated at $1.5 \times 10^{18} n/cm^2$	
		17.11		Nb_3Al			Neubauer
Pressure Coefficient	dT_c/dP	-0.7	10^{-5}°K/kg-cm^{-2}				
	T_c	7.0-12.0	°K	33	Nb_2Al, σ-tetragonal		Corenzwit
Critical Field	H_c	500	Oe	24.5		sintered T_c = 17.9°K	Raetz & Saur
				25		pressed granules, T_c = 17.6	
				29.3		pressed powder, 17.3	
Temperature Coefficient	dH_c/dT	370	Oe/°K	25	Nb_3Al	fused, T_c = 18.1	Meyer A
	dH_{c_1}/dT	240					
	dH_{c_2}/dT	22	kOe/°K				
Electrical Resistivity	ρ	105	$\mu\Omega$-cm	25	Nb_3Al	TEMP.(°C) 20	Samsonov & Sinelnikova A
		141				960	
		135		33	Nb_2Al	20	
		142				960	
		49		75	$NbAl_3$	20	Sinelnikova, Samsonov & Sinelnikova
		112				960	
Temperature Coefficient	$d\rho/dT$	0.0693	$\mu\Omega$-cm/°K	75		20-960	Samsonov & Sinelnikova A
Electronic Specific Heat	γ	31.81	mJ/mole °K^2		Nb_3Al	pressed high purity powders, reacted in solid state.	Willens et al.
		30.13				annealed at 700°C	
Debye Temperature		290	°K			non-annealed and annealed alloys	Willens et al.
Melting Point		2393	°K	25			Sinelnikova & Goralnik
		2463		33			
		1933		67			
Thermal Conductivity		0.293	W/cm °K		Nb_3Al	sintered	Sinelnikova
Thermal EMF		-22.84	μV/°C		Nb_3Al	sintered	Sinelnikova
Thermal Expansion Coefficient		11.2	10^{-6}/°C		Nb_3Al	Sintered	Sinelnikova
Microhardness		845	kg/mm^2		$NbAl_3$		Otto
Wave Velocity	V_L	5.64	10^5cm/sec		Nb_3Al	polycrystal, T_c=17.7°K, 4.2°K	Testardi et al.
		5.76				300. °K	

NIOBIUM ALUMINUM

PROPERTIES	SYMBOL	VALUE	UNIT	AT.% Al	CRYSTAL DATA	NOTES	TEMP.(°K)	REFERENCES
Penetration Depth		0.19	μ		Nb_3Al	fused, annealed-900°C	0	Meyer B
Magnetic Susceptibility	χ_{mol}	6.4 6.2	10^{-4} emu		Nb_3Al	solid state reacted, high purity powder	20 300	Willens et al.
	$\chi_{14°K}/\chi_{0°K}$	0.58 0.097				fused, annealed-900°C and milled 30 min. milled-6 hrs		Meyer A

NIOBIUM SCANDIUM

PROPERTIES	SYMBOL	VALUE	UNIT	AT.% Sc	Field	NOTES	TEMP.(°K)	REFERENCES
Transition Temperature	T_c	>4.2	°K	15		resistance meas.		Hake et al. A
Critical Current	J_c	1.7	10^4 Amp/cm^2		27 kG	ribbon, ॥-rolling plane	1.2	Hake et al. A
		1			20 kG		4.2	
		4.1	10^3 Amp/cm^2		20 kG	⊥-rolling plane	1.2	
		2.8			20		4.2	

NIOBIUM GALLIUM

PROPERTIES	SYMBOL	VALUE	UNIT	AT.% Ga	CRYSTAL DATA	NOTES	TEMP.(°K)	REFERENCES
Lattice Parameter	a_o	5.173	Å	24.7		annealed 2 hours 800°C		Kunz & Saur A
Transition Temperature	T_c	16.6 17.1	°K	10		arc-melted, resistivity meas. annealed 2 hours 800°C		Kunz & Saur A
		16.0 16.5		31		arc-melted, resistivity meas. annealed 2 hours 800°C		
		14.1 14.4		38		arc-melted, resistivity meas. annealed 2 hours 800°C		
		13.7 14.0		16		arc-melted, induction meas. annealed 2 hours 800°C		
		14.3 14.85		20-30 25		arc-melted, induction meas. annealed 2 hours 800°C		
		13.8 14.0		34		arc-melted, induction meas. annealed 2 hours 800°C		
		15.3		25		fusion		Testardi et al.
		15.6 16.1			Nb_3Ga	argon arc-melted annealed, a_o=5.170 Å		Otto
Critical Field	H_{c_2}	∿10 >28 10-15 0	kOe	<7 7-32 ∿35 ∿36			4.2	Guts et al. B
Hardness		885	kg/mm^2	25				Otto
Ductility		350 450-850	kg/mm^2	7-12 12-32		maximum		Guts et al. B
Wave Velocity Longitudinal	V_L	5.87	10^5 cm/sec	25			300	Otto

NIOBIUM YTTRIUM

PROPERTIES	SYMBOL	VALUE	UNIT	AT.% Y	CRYSTAL DATA	FIELD	NOTES	TEMP.(°K)	REFERENCES
Lattice Parameters	a_o	3.3010	Å	1			annealed 223 hrs. 1300°C		Taylor et al.
		3.2996		5					
		3.2993		25			annealed 10 min. 1200°C		
Transition Temperature	T_c	9.10	°K	1.32			arc-melted, cold-swagged. annealed in vacuo, 1 hr. 1200°C Yttrium is dispersed.		Koch & Love
Critical Field	H_{c_2}	4.3	kOe	1.32			arc-melted, cold swagged		Koch & Love
		3.4					annealed		
Critical Current	J_c	1.2	10^5 Amp/cm^2	1.32		2.15 kOe		4.2°K	Koch & Love
		2.1	10^4			3.44		4.2	
		5	10^2			4.3		4.2	

NIOBIUM INDIUM

PROPERTIES	SYMBOL	VALUE	UNIT	AT.% In	CRYSTAL DATA	FIELD	NOTES	TEMP.(°K)	REFERENCES
Lattice Parameters	a_o	5.3038	Å	25	Nb_3In		annealed		Otto
		5.303		25	β-W phase		annealed, P=40-70 kbar 1100°C		Banus et al., Killpatrick
Transition Temperature	T_c	9.28	°K	25			annealed		Otto
		9.2		25	β-W phase		annealed, P=40-70 kbar 1100°C		Banus et al.

NIOBIUM GADOLINIUM

PROPERTIES	SYMBOL	VALUE	UNIT	AT.% Gd	CRYSTAL DATA	FIELD	NOTES	TEMP.(°K)	REFERENCES
Transition Temperature	T_c	9.10	°K	0.51			arc-melted, cold swagged		Koch & Love
		9.10		5.3			annealed 1 hr. 1200°C Gadolinium is dispersed		
Critical Field	H_{c_2}	4.3	kOe	0.51			arc-melted		Koch & Love
		3.4					annealed 1 hr. 1200°C		
		4.3		5.3			arc-melted		
		3.6					annealed		
Critical Current	J_c	1.3	10^5 Amp/cm^2	0.51		2.15 kOe	arc-melted, non-annealed, a.c. susceptibility meas. at 4.2°K		Koch & Love
		4.0	10^2			4.3			
		1.8	10^5	5.31		2.15			
		2.8	10^2			4.3			

NIOBIUM CARBON

PROPERTIES	SYMBOL	VALUE	UNIT	AT.% C	CRYSTAL DATA	FIELD (kOe)	TEMP. (°K)	NOTES	REFERENCES
Lattice Parameters	a_0	3.117	Å	25.9	Nb+Nb$_2$S hexagonal				Hansen
	c_0	4.955							
	a_0	3.1194		32.4	Nb$_2$C+NbC hexagonal				Elliott
	c_0	4.9663							
	a_0	4.4244		36.4	NbC+Nb$_2$C cubic, (fcc)				Storms & Krikorian A
		4.4281		40.9	NbC, cubic				Storms & Krikorian B
		4.4559		45.3	cubic			powder, heated in vacuo, 2-24 hrs. 2000°C	Giorgi et al. A
		4.4693		49.1	cubic				Brown et al.
		4.470		49.9	cubic				Geballe et al.
		4.460		50	cubic				Avgustinik et al.
Transition Temperature	T_c	9.2	°K	28.5	hexagonal			arc melted	Giorgi et al.B
		9.18			Nb$_2$C, hex.			induction meas.	Hardy & Hulm
		<1.98		33.5	NbC$_{.51}$ hex.			heated to 2000°C	Giorgi et al. B
		<1.05		41.1	NbC, cubic			powder, heated in vacuo-2 to 24 hrs. 2000°C	Giorgi et al. C
		1.05		44.7	cubic				
		2.4		45.3	cubic			pressed powder, heated in helium 120 hrs. 1700°C	Toth et al.
		3.5		46.75	cubic			powder, heated in vacuo, 2-24 hrs. 2000°C	Giorgi et al. C
		4.2		46.83				susceptibility meas.	
		7.2		47.9					
		10.6		49.18					
		11.1		49.41					
		12.0		50				high density pressed powder, heated to 2600°C	Piper B
Critical Field	H_c	800	Gauss		NbC	9.		calc. values for experimental T_c=9°K	Fink et al. *
	H_{c_1}	120				9.			
	H_{c_2}	16.9	kGauss			9.			
Critical Current	J_c	400	Amp/cm^2	49.6		7	4.2	hot-pressed powder	Piper A
		1				12			
		750			NbC	12.6	4.2	H⊥J	Fink et al. *
		30				21.3	4.2		
Hall Coefficient	R_H	- 7.4	10^{-5}cm^3/C	41.5				sintered in vacuo 2300°C	Avgustinik et al.
		-13.3		44.7					
		-16.4		46.1					
		-11.2		47.6					
		- 7.0		49.0					
		-1.55	10^{-4}	50.				hot pressed powder 2600°C	Piper, B; Samsonov & Paderno
Electrical Resistivity	ρ	35.1	μΩ-cm		NbC		10.	normal state resistivity	Fink et al. *
		41.3		49.1			300	powder, sintered, 3000°C	Brenton et al.
Temperature Coefficient	dρ/dT	0.12	μΩ-cm/°K		NbC			sintered powder at 0-1200°C	Neshpor et al. B
		0.09			NbC			sintered powder at 30-2300°C	Samsonov & Sinelnikova B

PROPERTIES	SYMBOL	VALUE		UNIT	AT.% C	CRYSTAL DATA	TEMP.(°C)	NOTES	REFERENCES
Mobility		0.32		cm^2/V sec	41.5		20	pressed 20°C	Golikova et al.
		0.52			43.2			sintered in vacuo, 2400°C	B
		2.23			49				
		1.0				NbC	1870		Paderno et al.
Thermal EMF		8		μV/°K		$NbC_{0.76}$	200	sintered in vacuo	Golikova et al. A
		10				$NbC_{0.91}$	200		
		22					1000		
		38					2000		
		48				$NbC_{0.98}$	1800	sintered in vacuo	Golikova et al. B
		6				NbC	20		
Electronic Specific Heat		1.57		mJ/	32.4			pressed powder,	Toth et al.
		2.11		gr.at.-Nb°K^2	43.5			heated in He 120 hr. 1700°C	
		2.15			45.3				
		2.22			46.7				
		2.52			47.6				
		2.83			49.0				
Debye Temperature		464		°K	32.4		1°K		Toth et al.
		500			43.5				
		521			45.3		2.4		
		542			46.2		3.7		
		555			47.6		6.3		
		604			49.0		9.8		
		748			49.1		300		Brenton et al.
Thermal Conductivity	k_t k_1								
Total	k_t	6	5.5	mW/cm-°K	43.1		2°K	single crystal	Radosevich & Williams
		20	18				5		
Lattice	k_1	40	35				10		
		70	60				20		
		120	100				50		
		97			43.15		300	sintered in vacuo, 2400°C	Neshpor et al. A
Magnetic Susceptibility	$\chi_{gr.}$	2.5		10^{-7}emu		$NbC_{0.98}$	300		Borukhovich et al.
		2.04					22-300		
		0.40				$NbC_{0.70}$	20-300		Dubrovskaya et al.
Expansion Coefficient		6.35		10^{-6}/°C		$NbC_{0.98}$	25-2000°C	polycrystalline	Houska
		8.5				$NbC_{0.97}$	17-2500°C	polycrystalline	Brenton et al.
		6.65				NbC		sintered	Samsonov et al. B
		6.88							
		7.57							
Microhardness		1961		kg/mm^2		NbC			Samsonov & Paderno
Young's Modulus		4.92		10^{12}dynes/cm^2		$NbC_{0.97}$	25°C	hot-pressed, 3050°C corrected to 0-porosity	Brenton et al.
Bulk Modulus		2.68							Brenton et al.
Shear Modulus		1.84							Brenton et al.
Poisson's Ratio		0.22							Brenton et al.
Density		7.45		g/cm^3					Brenton et al.
Molar Volume		13.44		cm^3		$NbC_{0.964}$			Brown et al.
Melting Point		2335		°C				Eutetic, $NbC_{0.06}$-$NbC_{0.39}$	Storms & Krikorian B
		3080						Peritectic, $NbC_{0.5}$-$NbC_{0.56}$	
		3500				$NbC_{0.86}$			

PROPERTIES	SYMBOL	VALUE	UNIT	AT.% Si	CRYSTAL DATA Type	TEMP.(°C)	NOTES	REFERENCES
Lattice Parameters	a_o	4.211	Å	Nb_3Si	cubic, Cu_3Au		powder pellets, heated, 140 hrs. 1100°C	Galasso & Pyle *
	a_o	5.19		24.5	tetragonal		arc-melted in He	Deardorff et al.
	c_o	10.21						
	a_o	6.570		α-Nb_5Si_3	tetr. Cr_5B_3		pressed powder, heated in Ar 1100°C annealed, 12 hrs.1600°C	Kieffer et al. B
	c_o	11.884						
	a_o	5.077		β-Nb_5Si_3	tetr. Ni_3P			
	c_o	10.018						
	a_o	4.795		$NbSi_2$	hexagonal			Kieffer et al.,B Samsonov et al.A
	c_o	6.589						
Transition Temperature	T_c	1.5	°K	Nb_3Si				Galasso & Pyle *
		<4.2		24.5	tetragonal			Deardoff et al.
Electrical Resistivity	ρ	20	$\mu\Omega$-cm	1.64				Ulyanov & Tarasov
		52		$NbSi_2$		20	argon-arc melted	Samsonov et al. A
		94				800		
Temperature Coefficient	$d\rho/dT$	0.04	$\mu\Omega$-cm/°C					Samsonov et al. A
Hall Coefficient		-7.7	$10^{-5}cm^3$/C	$NbSi_2$		20		Neshpor & Samsonov
Mobility		1.08	cm^2/V sec	$NbSi_2$		20		Neshpor & Samsonov, Samsonov et al. A
Thermal EMF		10.35	μV/°C	66.1		20	argon-arc melted and annealed	Davisson et al.
		13.7		66.7				
		12.4		67.2				
		14.4		66.7				Neshpor & Samsonov
Density		7.35	gr/cm^3	24.5	tetragonal	20		Deardorff et al.
		5.45		$NbSi_2$			sintered 1600°C	Kieffer et al. B
		6.26		Nb_5Si_3			sintered 1600°C	Kieffer et al. B
Melting Point		1943	°C	17.9				Deardorff et al.
		1945		20				
		1945		25	Nb_3Si			
		2480			Nb_5Si_3			
		1930			$NbSi_2$			
Hardness		170	kg/cm^2	1				Ulyanov & Tarasov
		200		1.5				
		470-500		20	Nb_4Si			
		400-600		37.5	Nb_5Si_3			
		660-700		66.7	$NbSi_2$			

NIOBIUM TITANIUM

PROPERTIES	SYMBOL	VALUE	UNIT	AT.% Ti	CRYSTAL DATA	TEMP.($^\circ$K)	NOTES	REFERENCES
Lattice Parameters	a_o	3.287	$\overset{\circ}{A}$	17	cubic		arc-melted	Hansen et al.
		3.282		33			annealed, 200-600 hrs	
		3.280		45			600-750°C	
		3.279		56				
		3.277		65-75				
	a_o	3.13		80	orthorhombic		arc-melted, water-quenched	Hatt & Rivlin,
	b_o	4.87					annealed in vacuo, 24 hrs.	Baker & Sutton,
	c_o	4.64					1000°C	Brown et al.
	a_o	2.93		85-97.5	hexagonal		arc-melted and quenched,	Bucher &
	c_o	4.57					annealed in vacuo	Müller
Transition Temperature	T_c	9.0-9.2	$^\circ$K	0.5-12.5			arc-melted, high purity, cold-worked or annealed in vacuo, 30 min. 1800°C	Fietz & Webb A
		9.4		20-40			arc-melted, slow cooled	Hulm &
		9.5		50			or water-quenched	Blaugher
		9.0		55-60				
		9.2		37			annealed, 16 hr. 1300°C	Shapira &
		9.0		56				Neuringer, A
		9.7		67			arc-melted, cold-worked high defect-density	Dubeck & Setty
		8.9		78			annealed, 16 hr. 350°C	Salter et al.
		6.5		80			arc-melted, water quenched	Baker & Sutton
		7.0					annealed 1 hr. 330°C	
		8.5					10 330	
							100 330	
		6.5		78			cubic phase limit	Hulm &
		6.0		82			arc-melted, not annealed	Blaugher
		4.0		90				
		2.0		97.5				
		8.9		61			vapour-deposited	Coffey et al.
		8.66		61			deutron-irradiated	
		8.9		61			annealed	
Critical Field	H_{c_2}	2.16	kOe	0.5		0.13-9	arc-melted, high purity annealed in vacuo, 30 min. 1800°C	Fietz & Webb, A
		2.24		1.5				
		2.33		4.5				
		2.36		9.0				
		1.		3		4.2	wire	DeSorbo A
		33.2		5		1.2	arc-melted, cold-rolled $J=10$ Amp/cm^2	Berlincourt & Hake
		51.3		10				
		58		15				
		104		30				
		145		58-66				
		98		80				
		38		90				
		82		33		4	arc-melted, $J=1$-10 Amp/cm^2	Wernick et al.
	H_{c_2}	5.5		0.5		1.35	arc-melted, high purity annealed in vacuo, 30 min. 1800°C	Fietz & Webb, A B
		8.5		1.5				
		20		4.5				
		31		9.0				
		40		12		4.1		
		33		10		2	annealed	Kim et al.
		58.7		10		1.2	arc-melted, cold-rolled $J=10$ Amp-cm^2	Berlincourt & Hake
		64		15				
		112		30				
		129		40				

NIOBIUM TITANIUM

PROPERTIES	SYMBOL	VALUE	UNIT	AT.% Ti	CRYSTAL DATA	TEMP.(°K)	NOTES	REFERENCES
Critical Field	H_{c_2}	111	kOe	37		0	annealed, 16 hr. 1300°C	Shapira & Neuringer, A
		188		56				
		140		70		1.2	arc-melted, cold-rolled	Berlincourt & Hake
		128		79.3		1.2	arc-melted, annealed, cold-rolled	Vetrano & Boom
		50		80		4.2	arc-melted, quenched	Baker & Sutton
		15		80			annealed, 1 min. 330°C	
		45		80			annealed, 50 hr. 330°C	
		108		80		1.2	arc-melted, cold-rolled	Berlincourt & Hake
		44.8		90				
		19.1		95		0	arc-cast, cold rolled, annealed, 2 hr. 1250°C	Kroeger
		47.4		61		7	vapour-deposited	Coffey et al.
		41.4					deutron-irradiated	
		47.1					annealed at 20°C	

PROPERTIES	SYMBOL	VALUE	UNIT	AT.% Ti	Trans. Field	TEMP.(°K)	NOTES	REFERENCES
Critical Current	J_c	7.8	$10^3 Amp/cm^2$	10	22 kOe	4.2	annealed, 1 hr. 1100°C cold-worked	El Bindari & Litvak
		6.2			28			
		1.0		10	22	4.2	cold-rolled sheet, H⊥R.P.	Hake et al. A
		6.5		10	28	4.2	cold-rolled sheet, H∥R.P.	
		0.1		28-65			unrolled	Hake et al. B

		H⊥R.P.	H∥R.P.	AT.% Ti	CRYSTAL DATA	TEMP.(°K)	NOTES	REFERENCES
		0.1	3.5	28	30	4.2	cold-rolled sheet	Hake et al. B
		0.12	1.4	50				
		0.38	4.6	65				
		4.4	4.8	80		1.2		

		20°C	400°C	600°C	AT.% Ti	CRYSTAL DATA	TEMP.(°K)	NOTES	REFERENCES
		2	10	20	20	30	4.2	arc-melted, annealed, cold-drawn ,0.25 mm wire, annealed in vacuo for 1 hr. at temperatures as shown.	Rauch et al.
		1.5	80	5	50				

PROPERTIES	SYMBOL	VALUE	UNIT	AT.% Ti	CRYSTAL DATA	TEMP.(°K)	NOTES	REFERENCES
		45	$10^3 Amp/cm^2$	30	30	4.2	0.25 mm wire	Berlincourt
		23		39			0.125 mm wire	
		45		50			0.04 mm wire	
		70		60	30	4.2	1000 Å thick film	Edgecumbe et al.

PROPERTIES	SYMBOL	annealed	worked	UNIT	AT.% Ti	CRYSTAL DATA	TEMP.(°K)	NOTES	REFERENCES
Electrical Resistivity Normal State	ρ_n	0.07	0.6	μΩ-cm	0			arc-melted, high purity annealed in vacuo, 30 min. at 1800°C or cold-worked	Fietz & Webb, A
		0.5	1.0		0.5				
		1.4	1.9		1.5				
		3.5	4.6		4.5				
		8.6	8.9		9.0				
			12.3		12.5				
			19.0		15		1.2	arc-melted, cold rolled	Berlincourt & Hake
			30.6		30				
			79.4		70				
			85.7		70			wire	
			97.2		80			wire, reduction ratio=14.3	
			63.8		90			wire, reduction ratio= 9.7	
			99		67		10.5	arc-melted, cold rolled	Dubeck & Setty
			125		67		300		
			33		37		10	arc-melted	Shapira & Neuringer, A
			53		56			annealed, 16 hr. 1300°C	
			80		80		4	arc-melted, cold rolled	Baker & Sutton
			130					annealed 1 min. 330°C	
			80					annealed 100 hr. 330°C	

NIOBIUM TITANIUM

PROPERTIES	SYMBOL	VALUE	UNIT	AT.% Ti	CRYSTAL DATA	TEMP.(°K)	NOTES	REFERENCES
Electrical Resistivity		25	$\mu\Omega$-cm	15		300	arc-melted, cold rolled annealed 70 hrs. 1000°C water quenched	Ames & McQuillan
		35		25				
		70		50				
		90		65				
		110		80				
Electronic Specific Heat		7.8	mJ/mole-°K^2	0				Heiniger et al.
		7.6		0.5			Arc-melted, high purity annealed in vacuo 30 min. at 1800°C, or cold worked	Fietz & Loeb, A
		7.7		1.5				
		8.0		4.5				
		8.5		9				
		8.9		12.5				
		10.7		10, 25			cold-rolled strips	Lubell & Kroeger
		5.45		75		<17		Sukharevsky et al.
		4.3		96			arc-melted and quenched	Heiniger & Muller, Heiniger et al.
		3.32		100				
Electron Mean Free Path		3.27	Å	70				Hake B
Coherence Length		260	Å	70				Hake B
Thermal Conductivity		2.5	mW/cm °K	67		2	arc-melted, cold rolled	Dubeck & Setty
		2.7				3.4		
		11.4				4.2		
Debye Temperature		239	°K	75		0		Sukharevskii et al.
		340		96		0		Heiniger & Muller
Wave Velocity Longitudinal		5.04	10^5 cm/sec	32		4.2	annealed in vacuo, 16 hrs. 1300°C	Shapira & Neuringer, B
Shear		2.11						
Density		7.3	g/cm^3			300		Shapira & Neuringer, B
Vickers Hardness		220	kg/mm^2	80			arc-melted, quenched	Baker & Sutton
		410					annealed, 96 hr. 330°C	
Knoop Microhardness		330		20			arc-melted and annealed 12 hr. 1500°C aged at 300°C	Rauch et al.
		330		40			aged at 400°C	
		320		50			aged, 200-400°C	

PROPERTIES	annealed	non-annealed	AT.% Ti	wavelength(μ)	T_c	NOTES	REFERENCES
Refractive Index	2.06	2.30	33	1	10.3	annealed 7 hr. 1200°C	Leksina
	4.00	4.20		2			
	6.10	6.48		3			
	10.2	10.8		5			
	17.0	18.0		10			
	2.46	2.20	50	1	9.3		
	4.13	4.05		2			
	6.19	5.90		3			
	9.68	9.50		5			
	17.0	14.9		10			

NIOBIUM GERMANIUM

PROPERTIES	SYMBOL	VALUE	UNIT	AT.% Ge	CRYSTAL DATA	NOTES	REFERENCES
Lattice Parameters	a_o	5.1756	Å	13.75	$NbGe_{0.159}$ (β-W)	powders heated in vacuo 1 hr. at 1600°C--$Nb_3Ge_{0.55}Nb_{0.45}$	Carpenter & Searcy
		5.167		18.	$NbGe_{0.221}$ (β-W)	$Nb_3Ge_{0.72}Nb_{0.28}$	Reed et al. B
		5.166			Nb_3Ge (β-W)		Matthias et al. A
		5.168			Nb_3Ge (β-W)	Powders heat in vacuo, 1 hr. 1700°C	Carpenter & Searcy
		5.1743			Nb_3Ge (β-W)	arc-melted, annealed	Nevitt
		5.149		29	Nb_3Ge (β-W)	arc-cast, fast quenched	Matthias et al B
	a_o	5.370			$NbGe_{0.67}$		Nowotny et al.
	c_o	7.718			tetragonal, $D8_8$	0.4% Carbon	
	a_o	5.152			Nb_5Ge_3	vapour-phase deposition on platinum at 1200°C ($NbGe_{0.54}$)	Cherry et al., A Nowotny et al.
	c_o	10.148			tetragonal		
Transition Temperature	T_c	4.9	°K	13.75	$NbGe_{0.159}$	single-phase compact	Reed et al. B (p.29)
		5.4		18.	$NbGe_{0.221}$		
		6.9			Nb_3Ge		Matthias et al. A
		6.0			$Nb_{0.71}Ge_{0.29}$		Matthias et al. B
		>17.0			Nb_3Ge	arc-cast, fast quench	Matthias et al. A
		1.02			Nb_5Ge_3		Nowotny et al.
Debye Temperature		378	°K		Nb_3Ge	at 300°K	Testardi et al.

Wave Velocity

	4.2°K	300°K					
Longitudinal	5.84	5.78	10^5 cm/sec	Nb_3Ge			Testardi et al.
Shear	2.93	2.90					
Hardness		1135	kg/mm^2	Nb_3Ge			Testardi et al.

PROPERTY	SYMBOL	VALUE	UNIT	AT.% Zr	CRYSTAL DATA	TEMP. (°K)	FIELD	NOTES	REFERENCES
Lattice Parameter	a_o c_o	3.2330 5.1475	Å	100	α-Zr, hexagonal				Donnay
Density		6.501	gr/cm^3		α-Zr				
Phase Transformation		1143	°K		β-Zr, cubic				
Lattice Parameter	a_o	3.6162	Å	100	β-Zr	1252		stable above 1143°K	
	a_o	3.3005		0	Nb, cubic				
Density		8.55	gr/cm^3	0	Nb				

	>1600°C	700°C		500°C					
		β_1	β_2	β_1	AT.% Zr			NOTES	REFERENCES
Lattice Parameters	3.318				5			arc-melted and quenched at given temperatures. $\beta_1 = \beta_{Nb}$ $\beta_2 = \beta_{Zr}$	Knapton, B Rogers & Atkins, Shukovsky et al.
	3.330				10				
	3.359	3.317		3.314	20				
	3.391	3.323		3.315	30				
	3.419	3.323		3.317	40				
	3.447	3.322		3.316	50				
	3.447	3.328	3.509		55				
	3.483	3.325	3.505		60				
	3.503		3.503		70				
	3.533				80				

PROPERTY	SYMBOL	VALUE	UNIT	AT.% Zr	CRYSTAL DATA	TEMP. (°K)	FIELD	NOTES	REFERENCES
Transition Temperature	T_c	10.9	°K	10				arc-melted, homogenized, annealed in vacuo	Hulm & Blaugher
		10.8		15				argon arc-melted, vacuum annealed, 10 hrs. 1000°C	Masuda et al.
		10.5		20				strip, annealed 200 hrs. 570°C	Bychkov et al.
		10.35		15				0.38 mm wire	Evans & Erickson
		10.5		25					
		10.6		25				argon arc-melted, vacuum annealed	Masuda et al.
		11.5		27.5				arc-melted, high purity 0.2 mm wire, annealed 1100°C	Dietrich et al.
		10.2		33				0.19 mm wire	Evans & Erickson
		10.6		35					Masuda et al.
		10.9		40				arc-melted, annealed	Hulm & Blaugher
		9.8		50				0.249 mm wire	Evans & Erickson
		9.7		50					Masuda et al.
		8.3		75					
		8.0		80				0.2 mm wire	Dietrich et al.
		>4.6		96				arc-melted, quenched from β-phase, annealed, 40 days, 600°C	Oakridge Nat. Lab. (p.52)
		2.7		97					
		1.1		98					
		1.2		99					
Pressure Coefficient	dT_c/dP	22	10^{-6}°K/atm.	25				wire	Itskevich et al.

PROPERTY	SYMBOL	VALUE	UNIT	AT.% Zr	CRYSTAL DATA	TEMP. (°K)	TRANS. FIELD	NOTES	REFERENCES
Critical Field	H_{c2}	17	kOe	5		4.2	$10^3 A/cm^2$	arc-melted, annealed in vacuo, 0.25 mm wire	Ralls et al.
		35		10					
		44		15					
		69		25					
		71		29					
		78		33					
		87		38					
		96		65					
		55		83					
		6		94					
		95		55		4.2	$10 A/cm^2$	cold-worked, .25 mm wire, annealed 600°C annealed 650-1400°C	Shukovsky et al.
		64							
		85							
					turns/cm^2				
		32.5		25	900	1.3		solenoid, 0.25 mm wire	Boom & Roberts
		29				4.2			
		17.5			469	4.2			
		14.0			333				
		12.5			300				
		H⊥I H∥I							
		70 82		25		4.22		cold-drawn wire, 0.25 mm	Aron & Hitchcock A
		83 95				2.24			
		H_{c1} H_{c2}			Reduction Ratio				
		28.0 39.9		4.7	65.3	1.2	$10 A/cm^2$	argon arc-melted, annealed and quenched, cold-rolled to 0.05 mm thickness	Berlincourt & Hake
		63.0 78.0		11.8	8.2				
		102.4 110.1		25	12.5				
		113.0 120.8		50	9.4				
		126.3 129.7		67.3	wire				
		111.5 118.2		80	29.0				
		34.8 44.7		90	1.9				
		50		10		1.	$10 A/cm^2$	cold-worked or annealed	Jones et al.
		197		25					
		138		40					
		148		61					
		H⊥R.P. H∥R.P.							
Critical Current	J_c	1.1 2.7	$10^3 A/cm^2$	12		4.2	30 kOe	arc-melted, annealed cold-rolled to 2 mm thick strip.	Hake et al. A
		7.0 20		25					
		1.4 22		38					
		50 110		25		4.2	5 kOe	single crystal strip polycrystalline strip 10 mil wide	Walker & Fraser
		40 200		30	cut ∥ R.P.				
		30 600		30	cut 45°-R.P.				
		0.5 10	$10^4 A/cm^2$	12		4.2	23.5	cold-rolled strip	Borodich et al.
		1.0 20		20					
		0.2 4		50					
		3 20		12		4.2	23.5	annealed	Borodich et al.
		20 40		20					
		40 60		50					
		1	$10^3 A/cm^2$	75		4.2	10	arc-melted tapes, reduced 97% to 0.025 mm thickness	Treuting et al.
		30						annealed 2 hr. 600°C	
		6						annealed 4 hr. 600°C	
		5						annealed 8 hr. 600°C	
		1	$10^5 A/cm^2$	75		4.2	27	0.05 mm ribbons	Bychkov
		2		80				annealed 1 hr. 450-550°C	
Anisotropy	$J_{c\perp}/J_{c\parallel}$	0.8		80			17	max.	

PROPERTY	SYMBOL	VALUE		UNIT	AT.% Zr	CRYSTAL DATA	TEMP. (°K)	TRANS. FIELD (kOe)	NOTES	REFERENCES
		H⊥R.P.	H∥R.P.							
Critical Current	J_c	0.7	19.8	10^4A/cm^2	25		4.2	20	0.06 mm rolled strip	Chandresekhar et al.
			7.9						0.25 mm cold-drawn wire	
			30.						0.19 mm wire, annealed at 600°C	
		10 kOe	20 kOe							
		8	-	10^4A/cm^2	5		4.2		argon arc-melted,	Ralls et al.
		8	9		10				annealed in vacuo,	
		8	9		15				cold-drawn, 0.25 mm wire	
		6	5		25					
		30	20		29					
		10	7		33					
		6	3		38					
		3	2		55					
		-	10		65					
		0.7	0.2		75					
		1-10			75		4.2	0-70	0.26 mm wire, annealed at 400-600°C	Ruzicka et al.
		33			67		5		high purity arc-melted cold-drawn, 0.2 mm wire	Dietrich et al.
		70			25		4.2	10	cold-drawn, 0.25 mm wire	Aron & Hitchcock, B
		80							annealed at 600°C	
		10							annealed at 1200°C	
		10			33		4.2		cold-drawn, 0.25 mm wire	Wong
		40							annealed at 600°C	
		50							annealed at 1200°C	
		4.5			25		4.2	1	cold-drawn, 5 mm wire, annealed, 4 hr. 1250°C wound in 600 turn solenoid	Ullmaier
		80			25			20	cold-drawn, 0.25 mm wire	Sekula et al. A
		70			33					
		55			50					
		50			75					
		0.025	0.125 0.5 mm							
		20	3 1		25		4.2	20	cold-drawn wires of indicated size	Olsen et al.
		10	4 -		35					
						Annealing Temp. °C				
		1		10^4A/cm^2	55	-	4.2		high purity arc melted and homogenized,	Shukovsky et al.
		80				380			cold-drawn to 0.25 mm	
		1				475			wire, annealed in vacuo	
		2				575				
		2				675				
		0.7				800				
		0.2				900, 975				
		0.2				1075				
		1				1450				
		<10			28		4.2	10-20	cold-worked wire, swaged and annealed to 0.55 mm	Kneip et al.
		2			85		4.2	0-20	slab, annealed 166 hr. at 525°C	King et al.
		H⊥I	H∥I							
		40	20	10^4A/cm^2	50		4.2	15-20	sputtered planar film, 750 Å thick	Edgecumbe et al.
		35	100							

NIOBIUM ZIRCONIUM

PROPERTY	SYMBOL	VALUE			UNIT	AT.% Zr	CRYSTAL DATA	TEMP. (°K)	NOTES	REFERENCES
Coherence Length	ξ_o	222			Å	25				Hake B
Electron Mean Free Path	ℓ_o	3.35			Å	25				Hake B
Electrical Resistivity		5.5			$\mu\Omega$-cm	4.7		1.2	cold-rolled ribbons, 0.05 mm thick	Berlincourt & Hake
		12.0				11.8				
		32.6				25				
		53.6				50				
		68.8				67.3				
		93.6				89.8				
		10	**40**	**200**	**273°K**					
		16	17	25	31	15			0.38 mm wire	Evans & Erickson
		24	25	34	38	25			0.38 mm wire	
		31	32	40	50	33			0.19 mm wire	
		51	52	59	63	50			0.25 mm wire	
		47			$\mu\Omega$-cm	25			cold-worked, 0.25 mm wire	Chandresekhar et al.
		51							annealed in vacuo, 600°C	
		43							annealed in vacuo, 900°C	
		51.7				35			cold-drawn, 0.009 mm wire	Olsen et al.
		50.5							0.025 mm wire	
		48.5							0.25 mm wire	
		11.8				96		77	annealed at 600°C	Oakridge Lab. (p.52)
		20.6				97			heat-treated and quenched	
		8.8				98			annealed at 600°C	
		10.4				99			heat-treated and quenched	
		90				99		300		
Frequency Coefficient	$\sim f^{0.5}$ $\sim f^{1.7}$					25		4.2	0.25 mm wire, >1 kHz <0.5 kHz	Zar
Electronic Specific Heat		10.5			mJ/mole °K^2	15			argon arc-melted, annealed in vacuo 10 hrs. 1000°C	Masuda et al.
		10.2				25				
		9.5				35				
		8.5				50				
		6.7				75				
		3.98				97				Oakridge Lab. (p.52)
		2.93				98				
		3.05				99				
Debye Temperature		234			°K	15				Masuda et al.
		215				25				
		225				35				
		202				50				
		183				75				
		263				97				Oakridge Lab.
		274				98				
		275				99				
		2	**4**	**8°K**						
Thermal Conductivity		2	47	300	mW/cm°K	0.2			vacuum-annealed wires, 1 mm	Radhakrishna & Nielsen
		3	48	310		2				
		90	160	340		25				
		200			mW/cm°K	15-50		300	commercial wire	Fox & Reichenecker
		120				90.2		50°C	1.27 cm extruded wire	Powell & Tye
		140						200°C		
		20	**80**	**280°K**						
Thermal EMF		-0.5	2.7	-0.5	μV/°K	1			0.25 mm wire, high purity,	Weinberg & Schultz
		-0.3	2.5	-0.5		1.8			annealed 4 hr. 1000°C	
		-0.6	1.7	-0.4		3.8			All pass through 0-point at $\sim$240°K	

PROPERTY	SYMBOL	VALUE	UNIT	AT.% Zr	CRYSTAL DATA	TEMP. (°K)	NOTES	REFERENCES
Magnetic Susceptibility	χ_{mol}	200	10^{-6}emu	15		300	argon arc-melted annealed in vacuo 10 hrs. 1000°C	Masuda et al.
		194		25				
		186		35				
		178		50				
		163		75				
		192		25		293	argon arc-melted, 3 mm wires	Taniguchi et al.
		176		50				
		163		75				
Vickers Hardness		280	kg/mm^2	10		4.2	helium arc-melted powders cast in vacuo, 6 mm cylinders	Guts et al. A
		430		30				
		375 (min)		84				
		460 (max)		92				
Yield Strength		243	10^3 psi	33			cold-drawn, 0.5 mm wire	Wong
Tensile Strength		245	10^3 psi	33			cold-drawn, 0.5 mm wire	Wong
		305	10^3 psi	25			cold-drawn, 0.025 mm wire	Olsen et al.
		240					0.25 mm wire	
Young's Modulus		11.2	10^6 psi	25		4	0.25 mm wire	Sekula & Redman
		11.2				77		
		12.3				300		
		17.0		33		300	0.5 mm wire	
Thermal Expansion Coefficient		5.4	10^{-6}/°K	15-50		4-473	Commercial wire	Fox & Reichenecker

NIOBIUM TIN

PROPERTY	SYMBOL	VALUE	UNIT	AT.% Sn	CRYSTAL DATA	TEMP. (°K)	NOTES	REFERENCES
Lattice Parameters	a_o	3.306	Å	1-10			arc-melted, annealed 1200°C sintered at 2150°C	Van Vucht et al. Vieland
		5.285		12.5	β-W+Nb		sintered 120 hr. 1190°C	Vieland
		5.291		14	β-W+Nb		274 950°C	
		5.2814		18	β-W+Nb		3 1975°C	
		5.2880		20	β-W+Nb		4 1500°C	
		5.2839		21	β-W+Nb		3 1975°C	
		5.2866		24	β-W+Nb		3 1975°C	
		5.2903		25	β-W+Nb		120 1190°C	
		5.292		25		1000°C	sintered	Bachner & Gatos
		5.289		26.7			chemical deposition on 0.18 mm wire	Hanak et al. B
	a_o	5.656		40	Nb_3Sn_2, b.c. orthorhombic		decomposition of Nb_3Sn in liquid tin at 1000C; annealed at 880°C	Ellis & Wilhelm
	b_o	9.199						
	c_o	16.843						
	a_o	5.65		45.9	Nb_6Sn_5, orthorhombic		melting and diffusion, argon-annealed, 1200-1400°C	Charlesworth et al.
	b_o	9.17						
	c_o	16.93						
	a_o	5.655		60	Nb_2Sn_3		decomposition of Nb_3Sn in liquid tin at 1000C; annealed at 780°C	Ellis & Wilhelm
	b_o	9.860						
	c_o	19.152						
	a_o	5.644		67	$NbSn_2$, orthorhombic		diffusion, annealed at 870°C	Charlesworth et al.
	b_o	9.845						
	c_o	19.215						
Density		8.95	g/cm^3	25	Nb_3Sn			Van Vucht et al.
		8.6		45.9	Nb_6Sn_5			Van Vucht et al.
		8.08		67	$NbSn_2$			Van Vucht et al.
Transition Temperature	T_c	8.8	°K	0.5			arc-melted, homogenized at 1200°C	Van Vucht et al.
		6.8		2.5				
		6		8			powders sintered at 980°C annealed in vacuo at 1325-1800°C with increased tin loss	Courtney et al.
		7		11				
		8		16				
		10.5		17				
		17.5		21				
		7.5		15			powders sintered at 1500°C	Vieland
		13.0		20				
		17.6		24-25			powders sintered at 1975°C	Vieland

	as-melted	annealed				
	-	17	18	argon arc-melted, annealed 6 hrs. 1200°C	Kunz & Saur B
	17.2	17.9	20		
	17.6	18.0	24		
	17.5	17.9	35		
	17.4	17.8	50		

PROPERTY	VALUE	UNIT	CRYSTAL DATA	NOTES	REFERENCES
	18.08	°K	Nb_3Sn	fused at 1500°C, argon-annealed at 1250°C, Density = 6.1 g/cm^3	Wiedemann
Transition Interval	0.04	°K			

32

PROPERTY	SYMBOL	VALUE			UNIT	AT.% Sn	CRYSTAL DATA	TEMP. (°K)	NOTES	REFERENCES
Transition Temperature	T_c	T_c	ΔT_c							
Transition Interval	ΔT_c	17.9	2.2		°K	26.5			vapour deposited strip on ceramic, 0.03 mm thick, annealed in vacuo	Cody & Cullen, Cooper
		17.7	0.1			28.7				
		18.3	0.03			29.5				
		17.8	0.6			26.7			vapour deposited on 0.18 mm wire, 0.012 mm thickness	Hanak et al. B
		18.5	0.2			20			powder, sintered at 1200°C and annealed	Reed et al.
		18.1	0.07				Nb_3Sn		diffusion of Nb layers in Sn vapour or liquid, 9 hrs. 1200°C	Rinderer et al.
		18.328					Nb_3Sn		vapour deposited neutron-irradiated	Cooper
		18.128								
		18.17					Nb_3Sn		195-strand of 0.5 mm Nb wires, heated in vacuo with tin, 100 hrs.1100°C	Koch et al.
		vapour-dep.	sintered							
		14.6	18.2			20			vapour-deposition on wire or ribbon, or sintered	Cherry et al., p. 24 A
		17.8	17.9			24				
		2.07				45.9			diffusion or sintering;	Charlesworth et al., van Ooijen et al.
		2.65				67			also foils heated to 800°C	
Pressure Coefficient	dT_c/dP	-1.4			10^{-4}°K/atm		Nb_3Sn		diffusion layer on wire	Müller & Saur Lazarev et al.
		-0.25								
		-0.2							calc.	Ganguly & Sinha
		H_{c_1}	H_o	H_{c_2}						
Critical Field		0.19			kOe		Nb_3Sn	4	vapour-deposited, formed into powder pellets	Hecht
		0.18						7		
		0.14						10		
		0.03						17		
			>190				Nb_3Sn	4.2	sintered powders, 1000°C drawn to 0.25 mm wire	Hart et al., Cooper, Rosenblum et al.
			190-220						ribbons or wires	Montgomery & Sampson
Transverse			30				Nb_3Sn	15.7	diffusion of pure Nb-layers on 0.5 mm wires	Meyer & Wizgall
			25					16		
Longitudinal			30					16		
			20					16.5		
		0.825	2.475				Nb_3Sn	15.5	10μ particles, sintered 1200°C	Cherry et al. B
		0.210	0.825					17		
		0	0.200					17.5		
		0.350	1.350				Nb_3Sn	11.6	2μ particles, pressed but not sintered	Cherry et al. B
		0.225	0.800					13.5		
		0.200	0.520					14.7		
		0.100	0.200					16.2		
		0.250	1.100					16.3	vapour-deposited film	Cherry et al. B
		0.200	0.930					16.8		
			0.200				$NbSn_2$	2	foil heated to 800°C	van Ooijen et al.
			0.620					0		
			<0.6				Nb_6Sn_5	2.1	crystal formed at 1000°C	Enstrom
Temp. Coeff.	dH/dT	-0.44	-1.24		kOe/°K		Nb_3Sn		10μ particles	Cherry et al. B
		-0.074	-0.23						2μ particles	
		-0.27	-1.6						vapour-deposited film	
	dH_{c_2}/dT	-19					T_c=18.3°K		vapour-deposited	Cooper, Rosenblum et al.

NIOBIUM TIN

PROPERTY	SYMBOL	VALUE	UNIT	AT.% Sn	CRYSTAL DATA	TEMP. (°K)	FIELD (kG)	NOTES	REFERENCES
Critical Current	J_c	1.5	$10^5 A/cm^2$	3.2		4.2	<7.5	pressed, sintered in vacuo 2 hrs. 1000°C	Swartz & Rosner
		2.2		14.3					
		1.1		23.1					
		1.2		25					
		1.0		26.8					
		0.8		11		4.2	80	wire, annealed 16 hrs. at 1000°C	Buehler et al.
		2.0		20					
		1.0		25					
		0.3		32					
		0.03		61.4		2.5	0	sintered powders	Enstrom et al. A
		1.8				4.2	10	5μ hot pressed powders	Sahm et al.
		1.0			Nb_3Sn		20	sintered, 1200°C	
		0.4			Nb_3Sn	4.2	0-20	powder, sintered	Enstrom et al. B
		0.1				20-90		2 hrs. 900°C	

		bulk	wire						
		500	10^5	A/cm^2	Nb_3Sn	14	15	bulk data for powders fused at 2400°C; wire prepared by drawn-tube method	Kunzler
		100	8×10^4			15			
		10	2×10^4			16			
		0.1	10^3			17			

PROPERTY	SYMBOL	VALUE	UNIT	AT.% Sn	CRYSTAL DATA	TEMP. (°K)	FIELD (kG)	NOTES	REFERENCES
		3	$10^5 A/cm^2$			4.3	10	diffusion ribbon	Stauffer et al.
		0.25			Nb_3Sn	4.2	100	Niostan ribbon coil	Montgomery &
		0.8						Cryostrand wire coil	Sampson
		20			Nb_3Sn		2	vapour deposition on	Schindler &
		13					10	stainless steel ribbon 2.25 mm wide by 0.6 mm	Nyman

		Ag-	Cu-						
H⊥I, R.P.		12	9.5 $10^5 A/cm^2$		Nb_3Sn	4.2	41	vapour-deposited on 1 mm wide, stainless steel ribbon, 0.03 mm thick; Cu or Ag-plated	Hudson
H⊥I, R.P.		7.5	5.2						
H∥I, R.P.		23							

		H_t	H_1						
Transverse Field, H_t Longitudinal H_1		35	$10^5 A/cm^2$		Nb_3Sn	4.2	0	0.06 mm thick vapour-deposit on 0.66 mm wide ribbon	Cherry
		6					25		
		4					50		
		1	14				100		
			5.5				135		
			2				160		
		2-4				4.2	7.5	0.012 mm thick vapour-deposition on wire	Hanak et al. B
Electrical Resistivity	ρ_n	115	μΩ-cm	15		5		sintered powder, 1550°C	Courtney et al.
		80		20		8			
		12		24.5		<18		vapour-deposited	Cody et al.
		12						annealed, 2 hr. 900°C	
		12.5						48 hr. 900°C	
		5.0						48 hr. 1000°C	
		15		29.5		27		vapour-deposited, 0.4 mm thick strips, $T_c=18.3°K$	Cody & Cohen
		39				77			
		81				300			
		0			Nb_3Sn	<17		sintered powder, 1200°C	Cherry et al. B
		30				17			
		45				50			
		140				300			
		70		26.7		300		vapour-deposited on wire	Hanak et al. B
		30			Nb_3Sn	<18		vapour-deposited on ceramic	Woodard & Cody
		50				100			

PROPERTY	SYMBOL	VALUE	UNIT	AT.% Sn	CRYSTAL DATA	TEMP. (°K)	NOTES	REFERENCES
Electrical Resistivity		0	$\mu\Omega$-cm	67	$NbSn_2$	2.5	pressed-powder pellet, annealed in vacuo, 3 hr. 800°C	van Ooijen et al.
		25				3		
		60				3.5		
Specific Heat		6.7	mJ/g-at.°K		Nb_3Sn	43	high-purity, vapour grown crystal	Vieland et al. (p. 56)
		8.5				50		
		9.4				56		
Electronic Specific Heat		5.8	mJ/g-at.°K^2		Nb_4Sn	8		Gittleman et al.
		12.3		23.5		0	H=0-103 Gauss	Vieland et al.
		32			Nb_3Sn*	0	annealed, vapour-grown crystal	Vieland & Wicklund
		13.5			Nb_3Sn	0	H=52 kG, sintered powder	
		20				10	vapour-grown crystals	Gittleman et al.
		50				17	vapour-grown crystals, H=0 Gauss	Vieland & Wicklund
		55				18		
Thermal Conductivity		0.1	mW/cm °K	29.5		2.6	vapour-deposited strips, 30 mm x 10 m x 0.4 mm T_c=18.3°K	Cody & Cohen
		0.7				4		
		6.0				10		
		23.0				18		
		24.0				28		
Debye Temperature		228	°K		Nb_3Sn	0	sintered, 1700°C, specific heat meas.	Vieland & Wicklund
		262		23.5		0	specific heat meas.	Vieland et al.
		307		20		0	specific heat meas.	Vieland & Wicklund
		204			Nb_3Sn	50	elastic constant meas.	Keller & Hanak
		265				100		
		305				200		
		324				300		
Energy Gap	$2\epsilon_0/kT_c$	2.1 (1.0-2.8)	eV		Nb_3Sn	2-4.2	single crystals, T_c=18.3°K tunneling meas. on 5 separate planes; meas. also on vapour-deposited.	Hoffstein & Cohen
		3.56					vapour-deposited, calc. from thermal conductivity meas.	Cohen et al., Cody & Cohen
		3.77			Nb_3Sn	0	vapour-deposited, 100μ thick, IR meas., T_c=18.3°K also photoluminescence meas.	Bosomworth & Cullen, Fraas et al., *
		1.9					surface damaged by polishing	Bosomworth & Cullen
Mean Free Path	ℓ_0	10	$\overset{\circ}{A}$		Nb_3Sn		vapour deposition	Cherry et al. A Air Force Mater. Laboratory
		2.8			Nb_3Sn		calc.	Cody, Hake
Coherence Length	ξ	100	$\overset{\circ}{A}$		Nb_3Sn			Hoffstein & Cohen, Cherry et al.
Penetration Depth	λ_0	390	$\overset{\circ}{A}$		Nb_3Sn		vapour deposition	Cherry et al.
		2900					chemical deposition on platinum wires, critical field meas.	Cody
Magnetic Susceptibility	$\chi_{g.-at.}$	24	10^{-5}emu		Nb_3Sn	55	single crystal	Gittleman et al. Vieland et al.
		19				250		

* gr.at. volume = 11.1 cm^3, gr.at. = mole/4

PROPERTY	SYMBOL	VALUE			UNIT	AT.% Sn	CRYSTAL DATA	TEMP. (°K)	NOTES	REFERENCES
Elastic Moduli		0	35	300°K						
	C_{11}	1.646	1.679	2.538	10^{12} d/cm^2	24.5			ultrasonic meas. on single crystal	Keller & Hanak
	C_{12}	1.646	1.637	1.124						
	C_{44}	0.266	0.269	0.396						
Wave Velocity	▌[110]	5.022			10^5 cm/sec	24.5		300		Keller & Hanak
	▌[110]	4.654						4.2		
	▌[001]	1.737						4.2		
Thermal Expansion Coeff.		7.1			10^{-6}/°C		Nb$_3$Sn		chemical deposition on wire	Hanak et al.B
		9.8						25-700°C	chemical deposition on ceramic	Hanak et al.A
Modulus of Elasticity		9			10^6 psi		Nb$_3$Sn	300	vapour deposited	Air Force Mat. Lab.
Tensile Strength		24			10^3 psi					
Ductility		25			%		Nb$_3$Sn	300	0.75 mm wire core in a niobium sheath, enclosed in a Monel sheath	Shaheen
Breaking Point Load		48			lbs.					
Gr. At. Volume		11.1			cm^3		Nb$_3$Sn			Vieland & Wicklund

PROPERTY	SYMBOL	VALUE		UNIT	AT.% Hf	CRYSTAL DATA	TEMP. (°K)	FIELD	NOTES	REFERENCES
Lattice Parameter	a_o	3.44		$\overset{\circ}{A}$	70	cubic			helium arc-melted, annealed in vacuo, 48 hrs 1000°C quenched	Duwez
		3.42			56.5					
		3.40			53					
		3.38			45					
		3.36			34					
		3.34			23					
		3.345			20				arc-melted	De Sorbo B
		3.320			10					
Transition Temperature	T_c	9.4			25					Hake et al.A
		>4.2			75					Hake A
		H_{c1}	H_{c2}							
Critical Field		62.1	69.6	kG	12.5	cubic	1.2	$10A/cm^2$	arc-melted, quenched and cold-rolled	Berlincourt & Hake
		78.9	89.6		25					
		109.4	103.5		50					
		83.1	89.7		75					
		83	96		87.5	h.c.p.				
		17	>28		75	b.c.c.	4.2		arc-cast and cold-rolled	Hake, A
		1% Sn	10% Sn					Trans.		
Critical Current		100	100	Amps			4.2	5kOe	arc-melted, wire drawn to 0.03 in.	De Sorbo B
		4	65					10		
		0.06	50					15		
Critical Current Density	H∥R.P.	2.6		$10^3A/cm^2$	25		4.2	30	arc-melted, rolled strips	Hake et al., A
	H⊥R.P.	0.08								
		0.08							unrolled strip	
Electrical Resistivity		18.5		μΩ-cm	5				electron-beam melted, drawn to 1.5 mm wire, annealed 1400°C	Druzhinina et al.
		21.6			10					
		19.1			12.5				argon arc-melted, quenched and cold-rolled	Berlincourt & Hake
		36.3			25					
		69			50					
		124.4			75					
		53.2			87.5					
Work Function	φ	3.74		eV	1				arc-melted, quenched, annealed 24 hr. 2000°C	Dyubuya & Kultashev
		3.94			10-35					
		3.9			50					
		3.78			75					
Brinell Hardness		87		kg/mm^2	5				arc-melted, wire drawn to 1.5 mm. annealed 1400°C	Druzhinina et al.
		112			10					

PROPERTY	SYMBOL	VALUE	UNIT	AT.% N	CRYSTAL DATA	NOTES	REFERENCES
Lattice Parameters	a_o	3.303	$\mathring{A}$	<5	cubic	α-phase	Brauer & Jander
	a_o	3.050		15.9	hexagonal		
	c_o	4.946					
	a_o	3.056		29	hexagonal	β-phase	Schoenberg, Donnay, Brauer & Jander
	c_o	4.956					
	a_o	3.048		33			
	c_o	4.995					
	a_o	2.950		44.4		γ-phase, P$\bar{6}$m2	Schoenberg
	c_o	2.772					
				46.2		pure γ-phase	
	a_o	2.958		47.3			
	c_o	2.779					
	a_o	2.968		49.0		δ-phase, P6$_3$/mmc, Z2	
	c_o	5.535					
	a_o	2.952		50	hexagonal	ε-phase, P6$_3$/mmc, Z4	
	c_o	11.25					
				33-44		β+γ phases	
	a_o	6.873		55.5	tetragonal	nitridation of thin Nb films in ammonia	Terao *
	c_o	4.298					
	a_o	5.193		54.5	hexagonal		
	c_o	10.380					
	a_o	4.383		45.6		pressed Nb powder, nitrided	Geballe et al.
		4.391		47.6		at 60 hr. 1450°C	
Transition Temperature	T_c	9.12	°K	0.33		solubility limit of N in Nb, quenched from 1200°C	de Sorbo C
		9.24		1.64		wire nitridation	
		16.17			NbN	wire nitridation, 0.5 mm. 22 hr. P=65 atm. 1500°C	Sadagopan et al.
						wire nitridation, 0.5 mm.	Horn & Saur
		14.3		46.0		22 hr. 2 atm. 1300°C	
		15.1		46.5		22 4 1300	
		16.0		47.5		22 7 1300	
		16.4		48.1		22 10 1300	
		16.2		49.		22 50 1500	
		16.2		49.8		22 100 1500	
		15.7		51.		22 100 1650	
		13.68		45.6		pressed powder, nitridation 60 hr. 1450°C	Geballe et al.
		15.95			NbN	wire nitridation,	Neubauer
Press. Coeff.	dT_c/dP	+0.4	10^{-5}°K/kg-cm^{-2}			22 hr. 15 atm. 1500°C	
		+0.3				nitridation at 1300°C	
	T_c	16.5	°K	47.4		80-wire bundle, 0.72 mm, nitridation at	Horn & Sauer
	ΔT_c	broad				22 hr. 10 atm. 1300°C	
	T_c	16.5		48.1		22 100 1500	
	ΔT_c	sharp					
						25 mm wire nitridation,	Rögener
	T_c	13.6		43.2		23 hr. 0.5 atm. 1490°C	
		15.98		46.5		16 5 1490	
		14.70		48.7		92 32 1475	
		15.63		49.		22 80 1425	
		15.30		50.7		24 40 1340	
		14.93		51.2		24 42 1460	

PROPERTY	SYMBOL	VALUE	UNIT	AT.% N	CRYSTAL DATA	TEMP. (°K)	FIELD	NOTES	REFERENCES
Transition Temperature	T_c	13	°K		NbN			sputtered film, glass substrate, (111)-oriented 2250 and 5580 Å thick.	Saito et al.
		17.3			NbN			r.f. sputtered film, Nb at P_N=0.01 Torr, 760°C	Keskar et al.*
	ΔT_c	0.3	°K					260 Å grain size	
Critical Field	H_c	8	Oe	48.1		15.8		40-Nb wire bundle, 0.72 mm, nitridation: 22 hr. P=100 atm. 1500°C	Horn & Sauer
	H_{c_2}	132	kOe		NbN	4.2		T_c=16.1°K, wire nitrid. 10 100 1500	Hechler et al.
		153				0			
Temp. Coeff.	dH_{c_2}/dT	13.7	kOe/°K			16.1			

		powder	wire							
	H_{c_2}	0.4	0	kOe		NbN	16		powder nitridation, 3 hr, 1 atm. 1300°C	Cook et al.
		1.0					15.5			
		3.5					15.2		wire nitridation	Hechler et al.
		4.2					14.9		10 100 1500	
			25				14			
			40				13			
		9.2					13.4			
			50				12			
			120				6			
			132				4.2			
		∼200		kOe		NbN	4.2		sputtered films, 2550 and 5580 Å thick	Saito et al.
			1.48	kOe	0.23		4.2		wire nitridation, T_c=9.2°K	De Sorbo C
			140	Oe	15.9	tetr.	5.8		powder nitridation, 1450°C	Schroeder
			70				6.0			
			145		44.4		7.1			
			40				8.1			
Critical Current Density		1	10^3A/cm^2				4.2	120kOe	wire nitridation, 22 hr. 10 atm. 1400°C	Hechler et al.
		7						22 100 1500		
		7						70 100 1500		

		2550Å	5580Å							
		7		10^4A/cm^2		NbN	4.2	20	sputtered films	Saito et al.
			40					80		
		5	15					120		
		3	3					160		

		irrad.	non-irrad.							
Critical Current		15	30	Amps.			4.2	0	wire nitridation, 22 hr. 65 atm. 1500°C	Sadagopan et al.
		0.6	0.2					20	T_c=16.17°K, neutron	
		0.35	0.15					80	irradiation	
		6.0	10					120		
Electrical Resistivity		60	μΩ-cm		NbN	300		powder nitridation	Lvov et al.
		65		44.6	T_c=12.7°K			0.25 mm wire nitrid. 20 hr. 0.2 atm. 1350°C	Rögener
		91		46.5	15.98			16 5 1490	
		80		49	15.63			22 80 1425	
		220		51.2	14.93			24 42 1560	
		1.7		23		10		wire nitridation	De Sorbo C
		273-580		33	Nb$_2$N	300		reactive evaporation, 150-1500Å	Rairden
		120			NbN	18		sputtered film	Keskar et al. *
		390			NbN	300		sputtered film	Saito et al.

PROPERTY	SYMBOL	VALUE	UNIT	AT.% N	CRYSTAL DATA	TEMP. (°K)	FIELD	NOTES	REFERENCES
Hall Coefficient	R_H	-0.39	$10^{-4} cm^3/C$			0		powder nitridation, 1400°C	Piper B
		-0.13				300		3 hrs. 5 atm. 1500	Shulishova
Magnetic Susceptibility	χ_{mol}	30	$10^{-6} emu$			300			Shulishova
Specific Heat		163	mJ/mole °K	49.1		11	0	powder nitridation, 12 hr. 1300°C	Armstrong
		643				20	0		
		167				11	1 kOe		
Electronic Specific Heat		3.01	mJ/mole °K^2	45.5		0	0	powder nitridation 60 hr. 1450°C	Geballe et al.
		2.64		47.6		0			
		4.2		47.6		7			
		12.6		47.6		11			
		19.2		47.6		15.7			
		14		49.1		11		powder nitridation, 12 hr. 1300°C	Armstrong
		22.6				13			
		25.1				15			
Thermal Conductivity		33.5	mW/cm °K			300			Rassman & Merz
Melting Point		2573	°K		NbN				TPRC, v.5, p. 535
		2693			Nb_2N				
Thermal EMF		+0.6	μV/°K		NbN	300			Lvov et al.
		+2.8			NbN	300		arc-melted, annealed	Davisson et al.
Compressibility		2.1	$10^{-6} cm^2/kg$		NbN	83			Samsonov & Portnoy, p.35
Linear Expansion Coefficient		10.1	$10^{-6}/°C$			0-1200°C			
Density		8.33		32.4	Nb_2N	300		powder nitridation, 1300-1450°C	Brauer & Lander
		8.32		42.9	NbN (tetr.)				
		8.30		48.1	NbN (cubic)				
Energy Gap	2ε	4.50	meV		NbN	0		sputtered film, 2800Å $T_c=12.8°K$	Komenou et al.
	$2\varepsilon/kT_c$	4.08							
Debye Temperature		331	°K	45.5				powder nitridation, 60 hr. 1450°C	Geballe et al.
		307		47.6					

NIOBIUM PHOSPHORUS

PROPERTY	SYMBOL	VALUE	UNIT	AT.% P	CRYSTAL DATA	TEMP. (°K)	NOTES	REFERENCES
Lattice Parameters	a_o	3.325	Å		β-NbP		Tetragonal, $I4_122$ Z4	Donnay
	c_o							
	a_o	2.348		23.6	α-NbP		Tetragonal, I4/mmm Z2	
	c_o	5.69						
	a_o	3.340			β-NbP		Nb-powder heated in P-vapour	Scott et al.
	c_o	11.408					annealed, 1 week, 900°C	
	a_o	8.878			NbP$_2$		Monoclinic	Hulliger
	b_o	3.266						
	c_o	7.529						
	β	119°8'						
Density		6.15	g/cm^3		β-NbP			Donnay
		6.40			α-NbP			
		6.48			β-NbP		metal powder heated with CaP	Ripley
Microhardness		599	kg/mm^2					Ripley
Electrical Resistivity		0.4	mΩ-cm			76	powder sintered at 1730°C	Ripley
		1.7				293		
Magnetic Susceptibility	$\chi_{gr.}$	-0.45	10^{-6}emu			78		Scott et al.

NIOBIUM VANADIUM

PROPERTY	SYMBOL	VALUE	UNIT	AT.% V		TEMP. (°K)	NOTES	REFERENCES
Lattice Parameter	a_o	3.30	Å	10			argon arc-melted,	Wilhelm et al.
		3.27		20			annealed, 48 hr. 1075°C	
		3.24		30				
		3.185		50				
		3.12		70				
		3.06		90				
Transition Temperature	T_c	6.69		10			argon arc-melted, vacuum-	Ishikawa & Toth,
		4.70		25			annealed, 2 hr. 1330-1980°C	Sirota &
		3.85		50				Ovseichuk, A,
		3.87		75				Hulm & Blaugher
		4.25		90				
		5.52		100				
Critical Field	H_c	4	kOe	10		3.5	arc-melted, annealed	Sirota &
		3		20			10 hr. 1000°C	Ovseichuk, B
		1.5		30				
		2		40				
		6		60				
		8		80				
		9		90				
Critical Current	J_c	3	10^3A/cm^2	10-20		3.5	arc-melted, annealed	Sirota &
		1		30-40				Ovseichuk, B
		2		50				
		6		60				
		9		80-90				
Electronic Specific Heat		7.42	mJ/mol-°K^2	10			argon arc-melted, vacuum-	Ishikawa & Toth
		7.65		25			annealed, 2 hr. 1330-1980°C	
		8.16		50				
		8.80		75				
		9.23		90				
Debye Temperature		263	°K	10				Ishikawa & Toth
		278		25				
		305		50				
		319		75				
		361		90				

PROPERTY	SYMBOL	VALUE		UNIT	AT.% V	CRYSTAL DATA	TEMP. (°K)	NOTES	REFERENCES
Electrical Resistivity		4.2	300°K						
		1	25	$\mu\Omega$-cm	10			arc-melted, annealed	Sirota &
		19	38		30-50			10 hr. 1000°C	Ovseichuk, B
		15	35		60				
		4	20		90				
Magnetic Susceptibility	χ_{mol}	281.0		10^{-6}emu	10		297		Lam et al.
		252.5			50				
		235.0			80				
		225.0			90				
Melting Point		1815		°C	10			argon arc-melted, annealed	Wilhelm et al.
		1800			20				
		1812			30				
		1840			40				
		1890			50				
		1985			60				
		2090			70				
Microhardness		300		kg/mm^2	10			arc-melted, annealed	Sirota &
		450			30-50				Ovseichuk, B
		300			90				

PROPERTY	SYMBOL	VALUE	UNIT	AT.% As	CRYSTAL DATA	TEMP. (°K)	NOTES	REFERENCES
Lattice Parameters	a_o	9.357	Å	67	NbAs$_2$ monoclinic		single crystal, halide vapour-transport	Furuseth & Kjekshus, A, Saini et al.
	b_o	3.3823						
	c_o	7.792						
	a_o	3.4517			NbAs, tetragonal		thermal decomposition of NbAs$_2$ at 1100°C, also halide vapour-transport	Furuseth & Kjekshus, B, Saini et al.
	c_o	11.680						
Density		7.41	g/cm^3		monoclinic			Furuseth & Kjekshus, A
		7.93			tetragonal			Furuseth & Kjekshus, B
Magnetic Susceptibility	$\chi_{gr.}$	-0.44	10^{-6}emu		NbAs$_2$	100		Furuseth & Kjekshus, B
		-0.66			NbAs	100		

NIOBIUM ANTIMONY

PROPERTY	SYMBOL	VALUE	UNIT	AT.% Sb	CRYSTAL DATA	TEMP. (°K)	NOTES	REFERENCES
Lattice Parameters	a_o	5.263	Å		Nb_3Sb, cubic		β-W phase	Rothwarf et al.
	a_o	10.239			$NbSb_2$,		single crystal,	Furuseth &
	b_o	3.6319			monoclinic		β=120° 0.4'	Kjekshus, A & B
	c_o	8.333						
	a_o	5.2643			Nb_3Sb		thermal decomposition of $NbSb_2$ at 1000°C	Furuseth & Kjekshus, B
	a_o	10.314			Nb_5S_4		thermal decomposition at 830°C	Furuseth & Kjekshus, B
	c_o	3.5566						
Density		9.126	g/cm^3		Nb_3Sb			Rothwarf et al.
		8.22			$NbSb_2$			Furuseth & Kjekshus, A & B, Nevitt
		8.83			Nb_3Sb			
		8.17			Nb_5S_4			
Transition Temperature	T_c	4	°K		Nb_3Sb			Bachner et al.
Magnetic Susceptibility	χ_{gr}	0.35	10^{-6}emu		$NbSb_2$	100		Furuseth & Kjekshus, B
		0.53			Nb_5Sb_4			
		0.82			Nb_3Sb			
Thermal EMF		-3	μV/°K		Nb_3Sb			Bachner et al.

NIOBIUM TANTALUM

PROPERTY	SYMBOL	VALUE	UNIT	AT.% Ta	CRYSTAL DATA	TEMP. (°K)	NOTES	REFERENCES
Lattice Parameter	a_o	3.30	Å	0-100			a_{Nb}=3.3008Å, a_{Ta}=3.3026Å arc-melted, annealed-2000°C	Williams & Pechin
Transition Temperature	T_c	8.87	°K	0.87			arc-melted, 2 mm wires, annealed in vacuo, 1 wk, <m.p.	Ikushima & Mizusaki
		8.76		1.56				
		8.55		4.25				
		8.42		6.22				
		7.50		19.7				
		9.02		2			arc-melted, homogenized	Kimura et al.
		8.87		4				
		8.40		10				
		8.08		15				
		6.56		40				
		8.15		10			arc-melted, 3mm x 30mm rods	Ogasawara et al.
		7.51		20				
		6.25		40				
		5.31		60				
		4.65		80				
		6.25		50			6mm x 35mm rod	McConville & Serin
		6.3		50			4mm rod, electron beam melted	Vinen & Warren
		4.64		80			single crystal rod	Lowell
		4.55		95				
		5.40		60			pressed powder, zone melted in vacuo, single crystal rods and slabs	Sousa
		4.64		80				
		4.55		90				
		4.45		95				

PROPERTY	SYMBOL	VALUE H_c	H_{c_1}	H_{c_2}	UNIT	AT.% Ta	TEMP. (°K)	FIELD	NOTES	REFERENCES
Critical Field	H_c H_{c_1} H_{c_2}									
		2.05	1.75	4.4	kOe	0.87	0		arc-melted, 2 mm wires	Ikushima & Mizusaki
		2.03	1.70	4.5		1.56				
		1.98	1.37	5.3		4.25				
		1.85	1.34	3.18		5			0.51 mm wire	Gittleman et al.
		1.89	1.12	5.56		6.22				Ikushima & Mizusaki
		1.75	0.83	7.5		19.7				
		1.6				20			arc-melted	Corsan & Cook
				5.08		2			argon arc-melted and homogenized	Kimura et al.
				6.14		4				
				7.7		10				
				8.5		15				
		1.69	0.91	7.08		10			arc-melted rods	Ogasawara et al.
		1.37	0.55	8.73		30				
		1.27	0.48	8.6		40				
			0.6			50	4.2		high purity, 1.25 mm wire	LeBlanc & Griffiths
		1.32		6.72		50	0		4 mm electron-beam melted rod	Vinen & Warren
				5.4		50	4.2		0.51 mm wire	Cody
		1.16	0.375	8.6		50	0		0.5 mm wire, annealed in vacuo, 3 hrs. 2000°C	Gittleman et al.
		1.04	0.37	6.75		60	0		arc-melted rods	Ogasawara et al.
		0.83	0.33	4.26		80				
		0.84				85	0		arc-melted	Corsan & Cook
Temp. Coeff.	$(dH/dT)_{T_c}$	-0.365			kOe/°K	20	7.42		arc-melted	Corsan & Cook
		-0.362				30	6.80			
		-0.356				50	5.93			
		-0.349				70	5.15			
		-0.328				85	4.58			
		-0.406	-0.295	-0.690		5	9.13		0.5 mm wire	Gittleman et al.
		-0.335	-0.108	-2.1		50	6.94			
	(dH/dt)			-4.1	kOe	0.87			$t=T/T_c=1$	Ikushima & Mizusaki
				-4.6		1.56				
				-6.0		4.25				
				-6.9		6.22				
				-9.7		19.7				

Critical Current	J_c	as-drawn	annealed					Trans.		
			16	10^3A/cm^2	45	4.2		0 kG	cold-drawn, 0.4 mm wire annealed in vacuo, 25 hr. 1500°C	Heaton & Rose-Innes
		16						0.6		
		0	8					1		
			4					2-3.4		
		0	0					4-5		
			3	10^3A/cm^2	50	1.3			cold-rolled, 22 mm thick	Niessen & Staas
		13		10^2A/cm^2	45	4.2		0.5	cold-worked 3.3 mm rods	Ullmaier
		10						1.0		
		8						2.0		
		7						3.5		
		0						4.2		
Electrical Resistivity	ρ	16.8			μΩ-cm	2	293		arc-melted, homogenized, rolled, annealed in vacuo, 40 hr. 1400°C	Ulyanov & Tarasov
		17				4				
		18				6				
normal-state	ρ_n	2.9				10	4.2		arc-melted rods	Ogasawara et al.
		3.3				20				
		5.0				30				
		5.2				40				
		5.7				60				
		3.6				80				
		4.0				50			arc-melted rods	Viner & Warren

PROPERTY	SYMBOL	VALUE		UNIT	AT.% Ta	TEMP. (°K)	FIELD	NOTES	REFERENCES
Electrical Resistivity		0.219		$\mu\Omega$-cm	0.87	4.2	>4.5 kOe	arc-melted, 2 mm wires	Ikushima &
		0.373			1.56			annealed in vacuo,	Mizusaki
		0.843			4.25			1 week <m.p.	
		1.29			6.22				
		2.56			19.7				
		4.2	**300°K**						
		0.7	13.65		5			0.5 mm wire, drawn from	Cody et al.
		1.4	16.66		50			zone-refined rods	
Resistivity Ratio	$\rho_{300}/\rho_{4.2}$	**as-drawn**	**annealed**						
		19.5	32		5			0.5 mm wire, drawn from	Cody et al.
		11.9	15.7		50			zone-refined rods and annealed	
		80			0.87			2 mm wires, annealed,	Ikushima &
		46			1.56			1 week <m.p.	Mizusaki
		20			4.22				
		12			6.23				
		7.6			19.7				
		6.0			50			6 mm x 35 mm rod	McConville & Serin
		6.6			80			single crystal rod	McConville &
		13.5			95				Serin
Electronic Specific Heat		8.02		mJ/mol-°K^2	0.87			2 mm wires, annealed	Ikushima &
		7.93			1.56			1 week <m.p.	Mizusaki
		8.01			4.25				
		8.09			6.22				
		7.38			19.7				
		7.98			2-4			argon arc-melted and	Kimura et al.
		7.85			10			homogenized	
		7.35			15				
		6.95			30				
		6.70			40				
		7.25			20			arc-melted alloys	Corsan & Cook
		6.75			30				
		6.53			50				
		6.3			50			rod	McConville & Serin
		5.83			70			arc-melted alloys	Corsan & Cook
		5.99			85				
Debye Temperature		279			2			argon arc-melted	Kimura et al.
		272			4				
		270			6				
		271			10				
		260			15				
		259			30				
		260			40				
		261			50			arc-melted alloys	Corsan & Cook
		260			75				
		255			85				
Electron Mean Free Path	ξ	2700		°A	0.87			arc-melted, 2 mm drawn	Ikushima &
		1600			1.56			wire, annealed in vacuo	Mizusaki
		700			4.22			1 week <m.p.	
		460			6.23				
		230			19.7				
		230			10			arc-melted rods	Ogasawara et al.
		200			20				
		190			30				
		190			40				
		220			50				Vinen & Warren
		250			53			floating zone, rods	Calverley & Rose-Innes

PROPERTY	SYMBOL	VALUE	UNIT	AT.% Ta	TEMP. (°K)	NOTES	REFERENCES
Electron Mean Free Path		200 280	Å	60 80		arc-melted rods	Ogasawara et al.
Penetration Depth		410 420 460 500 630	Å	0.87 1.56 4.25 6.22 19.7	0	2 mm wires, annealed in vacuo 1 week <m.p.	Ikushima & Mizusaki
		610 670 840 890		10 20 30 40		arc-melted rods	Ogasawara et al.
		850		50		electron-beam melted rods	Vinen & Warren
		960 970		60 80			Ogasawara et al.
Melting Point		2490 2600 2760 2850	°C	10 30 50 90			Williams & Pechin
Thermal Conductivity		19 40 86 175	mW/cm °K	60 80 90 95	4.2	pressed powder zone-melted in vacuo, single crystal rods and slabs	Sousa

	80 at.%	95 at.%			TEMP. (°K)	NOTES	REFERENCES
		62			1.2	single crystal rods	Lowell
	45				1.4		
		30			1.8-2		
	19				2-2.5		
	24	57			3		
	36				3.8		
		82.5			4		

PROPERTY	SYMBOL	VALUE	UNIT	AT.% Ta	TEMP.	NOTES	REFERENCES
Thermal EMF		0 1.6 7.6	mV	70	760°C 1200 1900	high purity wire	Maykuth et al.
Spectral Emissivity	$\varepsilon_{0.65\mu}$	0.413 0.390 0.373 0.369		10 40 80 90	1400°K	3.2 mm x 120 mm tube, annealed, 30 min. 1500°C	Kotlyar & Voskobornik

Energy Gap	2Δ (mV)	$2\Delta/kT_c$		AT.% Ta	TEMP.	NOTES	REFERENCES
		3.91 3.95 3.89 3.75 4.0		0.87 1.56 4.25 6.22 19.7	0	arc-melted, 2 mm wires, annealed in vacuo	Ikushima & Mizusaki
		3.41 3.49		20 30, 50		arc-melted alloys	Corsan & Cook
	1.87	3.65		50		tunneling, Josephson junction	Barnes
		3.56 3.33		70 85			Ikushima & Mizusaki
	0.69	3.6		97			Barnes

PROPERTY	SYMBOL	VALUE	UNIT	AT.% Ta	TEMP. (°K)	NOTES	REFERENCES
Shear Stress		42 2-5	kg/mm^2	15-75	77 300	single crystal	Peters & Hendrickson
Poisson Ratio		0.345		95	2.6	single crystal cylinder	Brändli & Griessen
Young's Modulus		2120	10^9 dyne/cm^2				

NIOBIUM BISMUTH

PROPERTY	SYMBOL	VALUE	UNIT	AT.% Bi	CRYSTAL DATA	TEMP. (°K)	FIELD	NOTES	REFERENCES
Lattice Parameter	a_o	5.320	Å	25	Nb_3Bi			β-W phase formed at 32 kbars and 900°C	Killpatrick
Transition Temperature	T_c	3.05	°K						

NIOBIUM OXYGEN

PROPERTY	SYMBOL	VALUE	UNIT	AT.% O	CRYSTAL DATA	TEMP. (°K)	FIELD	NOTES	REFERENCES
Lattice Parameter	a_o	3.305	Å	1.2				oxygen gas dissolved into wires	Gebhardt & Rothenbacher
		3.310		2.5					
		3.3175		4.2					
Transition Temperature	T_c	8.78	°K	0.7				wire	De Sorbo, A
		8.04		1.52					
		7.80		1.80					
		7.04		2.6					
		5.84		3.83					
		9.02		6.43				solubility limit	

PROPERTY	SYMBOL	H_{c_1}	H_{c_2}	UNIT	AT.% O	CRYSTAL DATA	TEMP. (°K)	FIELD	NOTES	REFERENCES
Critical Field		0.6	7.0	kOe	0.7			4.2	wire, annealed, 1 hr. 100°C or cold-worked ribbon	De Sorbo, B
		1.1	9.67		1.52				wire	De Sorbo, A
		1.05	10.3		1.8					
		.84	11.6		2.6					
			5.4		0.2				wire	van Ooijen & van der Goot
			6.6		0.86					
			8.4		1.3					
		2.55	5.0		6.43					De Sorbo, A

PROPERTY	SYMBOL	VALUE	UNIT	AT.% O	CRYSTAL DATA	TEMP. (°K)	FIELD	NOTES	REFERENCES
Critical Current		2	$10^4 A/cm^2$	0.7			4.2	0.5 kOe	cold-worked ribbon or annealed wire	De Sorbo, B
		4-40		0.86			4.2	1.0	wire; varied metallurgical treatment	van Ooijen & van der Goot
Electrical Resistivity	ρ	0.82	μΩ-cm	0.20			4.2		0.8 mm wire	van Ooijen & van der Goot
		3.06		0.86						
		5.14		1.30						
Resistivity Ratio	$\rho_{293°K}/\rho_{4.2}$	17		0.20					0.8 mm wire	van Ooijen & van der Goot
		4.2		0.86						
		2.8		1.30						
Normal Resistivity	ρ_n	8.2		1.52			10		wire	De Sorbo, A
		13.7		2.60						
Electronic Specific Heat		7.37	$mJ/mol-°K^2$	.7					wire	De Sorbo, A
		6.86		1.52						
		6.86		1.80						
		6.70		2.60						
Magnetic Susceptibility	χ_g	2.25	10^{-6} emu	0.6					wire	Gebhardt & Rothenbacher
		2.17								
		2.11								
		2.02								
Hardness	H_{50}	150	kg/mm^2	0.6					wire	Gebhardt & Rothenbacher
		220		1.2						
		350		2.4						
		480		4.2						
		600		5.4						

NIOBIUM SULFUR

PROPERTY	SYMBOL	VALUE	UNIT	AT.% S	CRYSTAL DATA	TEMP. (°K)	NOTES	REFERENCES
Lattice Parameters	a_o	3.31	Å		NbS_2,		single crystal by iodine transport at 1075-1225°K	Jellinek et al.
	c_o	11.89			hexagonal			
Transition Temperature	T_c	6.1-6.3	°K		NbS_2		single crystal	Van Maaren & Schaeffer
Electrical Resistivity		150	$\mu\Omega$-cm		NbS_2	298	single crystal	Van Maaren & Harland
		50				77		
Electronic Specific Heat		10.7	mJ/mol-°K^2					Van Maaren & Harland
Debye Temperature		265	°K					Van Maaren & Harland

NIOBIUM CHROMIUM

PROPERTY	SYMBOL	VALUE	UNIT	AT.% Cr	CRYSTAL DATA	TEMP. (°K)	NOTES	REFERENCES
Lattice Parameter	a_o	3.280	Å	1			argon arc-melted	Ulyanov & Tarasov
		3.272		4				
		3.248		10			Nb-solubility limit	Goldschmidt & Brandt
		2.885		96			Cr-solubility limit	
Transition Temperature	T_c	6.6	°K	5			arc-melted	Hulm & Blaugher
		4.5		10				
		4.5		10			annealed 2200°C	
Electrical Resistivity		17.5	$\mu\Omega$-cm	2.5				Ulyanov & Tarasov
		19.2		4.0				
Hardness		185	kg/mm^2	2.5				Ulyanov & Tarasov
		220		4.0				

NIOBIUM SELENIUM

PROPERTY	SYMBOL	VALUE	UNIT	AT.% Se	CRYSTAL DATA	TEMP. (°K)	FIELD	NOTES	REFERENCES
Lattice Parameters	a_o	3.45	Å		$NbSe_2$,			low-temperature form	Beerntsen et al., Revolinsky et al.
	c_o	12.54			hexagonal				
	a_o	3.44						high-temperature form	
	c_o	25.24						single crystals of both forms are prepared by iodine transport	Brixner
Density		6.22	g/cm^3					low-temperature form	Brixner
		6.46						high-temperature form	
Transition Temperature	T_c	5.15-5.62	°K		$NbSe_2$				Revolinsky et al.
Critical Field	H_{c_1}	0.2	kOe				4.2		Beerntsen et al.
	H_{c_2}	7							
Critical Current Density	J_c	60	$10^2 A/cm^2$		$NbSe_2$	4.2	0.2 kOe	single crystals, $\perp$ c	Beerntsen et al.
		5					7.0		
		40					0.2	‖ c	
		0.6					7.0		
Electrical Resistivity	ρ	16	$\mu\Omega$-cm		$NbSe_2$	7			Beerntsen et al.
		224				300			
Temp. Coeff.	$d\rho/dT$	0.65	$\mu\Omega$-cm/°K			<150-300		single crystal	Bachmann et al.
Electronic Specific Heat		20.5	mJ/mol-$°K^2$		$NbSe_2$				Van Maaren & Harland
		15						single crystal, calc. from reflectivity meas.	Bachmann et al. *
Debye Temp.		210	°K						Van Maaren & Harland
Magnetic Susceptibility	χ_g	0.7	10^{-6}emu			100			Selte & Kjekshus
Thermal Conductivity		0.021	W/cm °K			300			Brixner
Refractive Index	n	wavelength (μ)							
	1.755	0.545			$NbSe_2$			single crystals, iodine vapour transport	Myers & Montet
	1.784	0.590							

49

NIOBIUM MOLYBDENUM

PROPERTY	SYMBOL	VALUE	UNIT	AT.% Mo	TEMP. (°K)	FIELD (kOe)	NOTES	REFERENCES
Lattice Parameter	a_o	3.154	Å	10			Niobium and Molybdenum form a continuous solid solution, all b.c. cubic phase	Goldschmidt & Brandt
		3.166		20				
		3.210		50				
		3.261		80				
		3.282		90				
Transition Temperature	T_c	7.84	°K	5			zone-melted, single crystal rod	French et al.
		6.38		10				
		5.30		15				
		4.22		20				
		5.3		10				Morin & Maita, Brandt & Ginzburg
		3.4		25				
		0.76		38				
		0.50		40				
		0.31		42				
		0.181		43				Hein et al., B & C
		0.158		44				
		0.148		45				
		0.108		48				
		<0.03		60				
		0.30		90				
		0.50		95				
Pressure Coefficient	dT_c/dP	-0.4	10^{-5}°K/atm	25			arc-melted cylinders	Brandt & Ginzburg

Critical Field

PROPERTY	H_c	H_{c_1}	H_{c_2}	UNIT	AT.% Mo	TEMP. (°K)	FIELD (kOe)	NOTES	REFERENCES
Critical Field			<10	kOe	4.8	4.2		arc-melted, annealed in vacuo 16 hrs. 1600°C	Cappelletti et al.
	1.07	0.49	4.27		5	4.17		zone-refined single crystal	French et al.
	0.79	0.29	4.14		10	3.78			
	0.46	0.16	2.47		15	3.77			
	0.50	0.15	3.0		20	2.39			
			3		15	4.18		field transverse to slab	Lowell et al.

Critical Current

PROPERTY	I⊥H	I∥H	UNIT	AT.% Mo	TEMP. (°K)	FIELD (kOe)	NOTES	REFERENCES
Critical Current	200		A/cm^2	5	4.17	0.4	zone-refined single crystal	French et al.
	10					1		
	2					4		
	0.7					5		
		200				4.3		
		1				6.5		

PROPERTY	SYMBOL	VALUE	UNIT	AT.% Mo	TEMP. (°K)	FIELD (kOe)	NOTES	REFERENCES
Electrical Resistivity	ρ	3.07	μΩ-cm	5	4			French et al,
		4.72		10				
		6.05		15				
		7.10		20				
		~10		30-40				
		~20		60-95				
Resistivity Ratio	ρ_{300}/ρ_4	6.69		5			single crystals, cylinders or thin slabs	French et al., Sousa
		4.11		10				
		3.15		15				
		2.69		20				
Electronic Specific Heat		5.76	mJ/mol-°K^2	10				Morin & Maita
		4.52		25				
		3.26		38				
		2.87		40				Blaugher et al.
		2.68		42				Morin & Maita
		2.01		50				
		2.87		40			arc-melted, high purity powders, annealed in vacuo 20 hrs. 2000°C	Veal et al.
		1.62		60				
		1.46		70				
		1.49		80				
		1.67		90				

PROPERTY	SYMBOL	VALUE	UNIT	AT.% Mo	TEMP. (°K)	NOTES	REFERENCES
Electronic Specific Heat		1.62	mJ/mol-°K^2	60			Blaugher et al.
		1.67		80			
		1.7		80	3		Hardy & Miller
		2.2			5		
		2.8			7		
		4.0			10		

Thermal Conductivity

		2.5	3	4°K			
			40	60	mW/cm°K	2	single crystal, 3 mm cylinders or thin slabs — Sousa
		15	21			15	
		8	19			20	

PROPERTY	SYMBOL	VALUE	UNIT	AT.% Mo	TEMP. (°K)	NOTES	REFERENCES
Debye Temperature		247	°K	4.8			Cappelletti et al.
		260, 290		10		arc-melted and vacuum annealed; double values taken below and above change in slope of C/T	Morin & Maita
		290, 320		25			
		320, 330		38			
		340		40, 42			
		371.1		40		arc-melted, high purity powders, annealed, 20 hr. 2000°C	Veal et al.
		380		50			Morin & Maita
		429.4		60			Veal et al.
		441.9		70, 80			
		461.3		80			Morin & Maita
		487.4		90			Veal et al.
Magnetic Susceptibility	χ_{mol}	200	10^{-6} emu	10		arc-melted	Masuda & Nishioka
		188		20			
		172		30			
		146		40			
		124		50			
		88		70			
		78		80			
Energy Gap	2Δ	2.49	meV	4.8		optical meas. on arc-melted slabs, $T_c=8.0$°K	Cappelletti et al.
	$2\Delta/kT_c$	3.61					
Fermi Level		7.02	eV	35		phonon dispersion meas. on single crystals	Powell et al.
		8.24		75			
Spectral Emissivity	$\epsilon_{0.665\mu}$	0.35		0-100	2000°C		Thomas, A
Refractive Index	$n_{0.715\mu}$	4.75		10-20			Thomas, B
		4.1		60			
		2.0		70			
		4.1		90			
Hardness		160	kg/mm^2	5	293		Ulyanov & Tarasov
		185		8			
		200		10			

Shear Stress

		90	300°K				
		35	7	kg/mm^2	5	single crystal	Peters & Hendrickson
		50	18		11		
		>100	35		21		

NIOBIUM TELLURIUM

PROPERTY	SYMBOL	VALUE	UNIT	AT.% Te	CRYSTAL DATA	TEMP. (°K)	NOTES	REFERENCES
Lattice Parameters	a_o	10.904	Å		$NbTe_2$, hexagonal			Brixner
	c_o	19.888						
Density		7.62	g/cm^3					Brixner
Transition Temperature	T_c	0.6	°K		$NbTe_2$			Brixner
		0.74						Van Maaren & Harland
Electrical Resistivity		77	$\mu\Omega$-cm		$NbTe_2$	77	sintered bars	Brixner
		260				300		
		30				77	single crystals	Van Maaren & Harland
		130				300		
Electronic Specific Heat		3	mJ/mol-°K^2					Van Maaren & Harland
Thermal Conductivity		0.019	W/cm-°K		$NbTe_2$	300	sintered bar	Brixner
Magnetic Susceptibility		0			$NbTe_2$	300	zero or slightly negative to 725°K	Selte & Kjekshus

NIOBIUM TUNGSTEN

PROPERTY	SYMBOL	VALUE	UNIT	AT.% W	CRYSTAL DATA	TEMP. (°K)	NOTES	REFERENCES
Lattice Parameters	a_o	3.280	Å	2.5				Ulyanov & Tarasov
		3.272		5				
		3.268		8				
		3.26		20				De Sorbo, B
		3.25		35			solid solutions in all proportions	Kieffer et al. A
		3.2		60				
Transition Temperature	T_c	7.1		10				Ralls & Wulff, Hulm & Blaugher
		5.8		15				
		4.7		20				
		3.6		25				
		1.5		35				
Critical Field	H_{c_1}	1.4	kOe	9.2		3.15	heavily cold-worked rod	Hulm & Blaugher
	H_{c_2}	6						
Electrical Resistivity		4	$\mu\Omega$-cm	5		300		De Sorbo, B, Ulyanov & Tarasov
		22		8				
		13		20			arc-cast	Kieffer et al. A
		23		30				
		35		50				
		28		80				
		22		90				
Vickers Hardness		410	kg/mm^2	20			arc-cast	Kieffer et al. A
		480		40, 50				
		600		60				
		430		80				
		230		90				

NIOBIUM TECHNETIUM

PROPERTY	SYMBOL	VALUE	UNIT	AT.% Tc	CRYSTAL DATA	TEMP. (°K)	NOTES	REFERENCES
Lattice Parameters	a_o	3.276	Å	10	b.c.c.			Van Ostenburg et al.
		3.244		20				
		3.217		30				
		3.192		40				
		3.170		50				
		3.159		60				
		9.625		75	$NbTc_3$, cubic, α-Mn		argon arc-melted	Compton et al.
		9.547		85				Van Ostenburg et al.
Transition Temperature	T_c	10.5	°K		$NbTc_3$		argon arc-melted	Compton et al.
Magnetic Susceptibility	χ_g	195.8	10^{-6}emu	5		298	argon arc-melted, annealed 1 week 700°C	Van Ostenburg et al.
		108.9		30				
		63.3		50				
		136.5		75				
		114.8		85				
		138.3		97				

NIOBIUM RHENIUM

PROPERTY	SYMBOL	VALUE	UNIT	AT.% Re	CRYSTAL DATA	TEMP. (°K)	NOTES	REFERENCES
Lattice Parameter	a_o	3.25	Å	20			arc-melted, solid solution	Levesque et al.
		3.22		33				
		3.20		40				
		9.670		63	cubic, α-Mn		stable from 60-82% Re	Greenfield & Beck
		9.781		60			arc-melted	Bucher et al. C
Density		15.3	g/cm^3	60				Bucher et al. C
Lattice Parameter	a_o	9.72	Å	50	σ-phase			Greenfield & Beck
	c_o	5.07						
Density		14.4	g/cm^3					Greenfield & Beck
Transition Temperature	T_c	2-3.8	°K	50			arc-melted	Bucher et al. C
		2.5		60				
		4.8		56				Alekseevski & Mikhailov, Bucher et al.
		2.45		62				
		5.60		71				Bucher et al. F
		8.83		80				
		8.89		82				Blaugher & Hulm
Electronic Specific Heat		2.35	$mJ/mol\text{-}°K^2$	62				Bucher et al. F
		3.58		71				
		5.00		80				
Debye Temperature		326	°K	62		0	arc-melted	Bucher et al. F
		322		71				
		316		80				
Electrical Resistivity		17	μΩ-cm	1				Ulyanov & Tarasov
		17.5		2.5				
Magnetic Susceptibility	χ_g	98	10^{-6}emu	62				Bucher et al. C
		61		50				

NIOBIUM COBALT

PROPERTY	SYMBOL	VALUE	UNIT	AT.% Co	CRYSTAL DATA	TEMP. (°K)	NOTES	REFERENCES
Lattice Parameter	a_o	3.5442	Å	100				Donnay
Transition Temperature	T_c	8.5 7.0 6.15	°K	0.8 4.5 7.5			melted in vacuo	Krivko
g-Factor		2.090 2.065 2.051		0.8 4.5 7.5		4.2		Krivko

NIOBIUM RHODIUM

PROPERTY	SYMBOL	VALUE	UNIT	AT.% Rh	CRYSTAL DATA	TEMP. (°K)	NOTES	REFERENCES
Lattice Parameters	a_o	3.265 3.245	Å	12 18	α-Nb			Ritter et al
	a_o c_o	9.869 5.106		29.7	σ-phase			
	a_o c_o	9.80 5.07		40	σ-phase		electron-beam melted	Bucher et al. C
	a_o c_o	4.019 3.809		51.3	tetragonal			Ritter et al.
	a_o b_o c_o	2.827 4.770 13.587		55.9	orthorhombic			
	a_o b_o c_o	2.813 4.818 4.510		58.8	orthorhombic			
	a_o b_o c_o	2.806 4.772 20.25		62.3	monoclinic			
Density		9.9	g/cm^3	40			electron-beam melted	Bucher et al. C
Transition Temperature	T_c	4.21	°K	40				Bucher et al. F
		3.76 3.07 3.00 2.7		51.5-52.5 54 -55.5 56 -58.5 59.5-63.5			arc-melted, high purity	Sadagopan & Gatos
Electronic Specific Heat		3.52	mJ/mol-°K^2	40				Bucher et al. F
Debye Temperature		329	°K	40		0		Bucher et al. F
Magnetic Susceptibility	χ_g	8.2	10^{-7}emu	40				Bucher et al. C

PROPERTY	SYMBOL	VALUE	UNIT	AT.% Ru	CRYSTAL DATA	TEMP. (°K)	NOTES	REFERENCES
Lattice Parameters	a_o	3.27 3.230	°A	10 20	b.c.c.			Bender et al., Hurley & Brophy
	a_o c_o	3.147 3.218		40	b.c. tetragonal			
Transition Temperature	T_c	4.20 2.8 <1	°K	7.5 10 20-30			electron-beam melted in vacuo, high temperature annealed	Bucher et al., A
Electronic Specific Heat		4.9 4.1 1.9 1.6 1.86	mJ/mol-°K^2	7.5 10 20 30 38				Bucher et al., A
Debye Temperature		290 304 330 372 405 410	°K	7.5 10 20 30 38 40				Bucher et al., A

NIOBIUM PALLADIUM

PROPERTY	SYMBOL	VALUE	UNIT	AT.% Pd	CRYSTAL DATA	TEMP. (°K)	NOTES	REFERENCES
Lattice Parameters	a_o	9.77	Å	40	α-Mn		electron-beam melted in vacuo	Bucher et al. C
Density		10.2	g/cm^3	40	α-Mn			Bucher et al. C
Transition Temperature	T_c	2.47	°K	40	α-Mn			Bucher et al. C
		1.6		40	σ-phase			Bucher et al. F
Electrical Resistivity		24	μΩ-cm	2		300		Ulyanov & Tarasov
Electronic Specific Heat		2.66	mJ/mol-°K^2	40	σ-phase			Bucher et al. F
Debye Temperature		333	°K	40	α-Mn			Bucher et al. B
		293		40	σ-phase			Bucher et al. F
Magnetic Susceptibility	χ_g	5	10^{-7} emu	40	α-Mn			Bucher et al. C
Hardness		130	kg/mm^2	2		300		Ulyanov & Tarasov

NIOBIUM OSMIUM

PROPERTY	SYMBOL	VALUE	UNIT	AT.% Os	CRYSTAL DATA	TEMP. (°K)	NOTES	REFERENCES
Lattice Parameters	a_o	5.1359	Å	25			argon arc-melted, annealed and quenched	Zegler
	a_o	9.844		40	σ-phase, tetragonal			Knapton, A Matthias et al. C
	c_o	5.056						
	a_o	9.760		50	α-Mn, cubic			Knapton, A
		9.72		55				Dwight,
		9.685		60				Matthias et al. C
		9.635		67				
		9.655		67				
Density		11.59	g/cm^3	25				Geller et al.
		13.4		40	σ-phase			Bucher et al. C
		14.7		40	α-Mn			
Transition Temperature	T_c	0.94	°K	25				Hein et al. A
		<1.7		25			argon arc-melted	Zegler
		1.78		40				Matthias et al. C
		1.89		40				Bucher et al. F
		2.86		50				Bucher et al. C
		2.52		67				Matthias et al. C
Electronic Specific Heat		2.7	mJ/mol-°K^2	40				Bucher et al. F
Debye Temperature		310	°K	40				Bucher et al. F
Critical Field Temp. Coeff. $(dH_{c_2}/dT)_{T_c}$		1.26	Oe/°K	25				Hein et al. A
Magnetic Susceptibility	χ_{mol}	7.4	10^{-7} emu	40				Bucher et al. C
		6.0		50				

NIOBIUM IRIDIUM

PROPERTY	SYMBOL	VALUE		UNIT	AT.% Ir	CRYSTAL DATA	TEMP. (°K)	NOTES	REFERENCES
Lattice Parameters	a_o	3.262		Å	15	β-W, cubic		as melted	Geller et al.
		5.139			25				
		5.131			25				Zegler
		a_o	c_o						
		9.888	5.072		35	σ-phase, tetragonal		formed at m.p.	Knapton A
		9.892	5.072		34			arc-melted, annealed in vacuo, 1600°C	Koch & Scarbrough
		9.869	5.062		37				
		9.842	5.045		40				
Density		11.52		g/cm^3	25				Geller et al., Zegler
		13.1			37				Bucher et al. C
Transition Temperature	T_c	1.7		°K	25				Zegler
		2.25			34				Koch & Scarbrough
		2.4			37				Bucher et al. C
		2.23			37				Koch & Scarbrough
		2.16			38.5				
		2.20			40				
		7.70			40	σ-phase			
		9.8			40	σ-phase			Matthias et al. C
		4.75			52-53			high-purity, arc-melted	Sadagopan & Gatos
		4.6			54.5-59.5				
		0.17			92.5				Andres & Jensen
Electronic Specific Heat		3.037		mJ/mol-°K^2	34				Koch & Scarbrough
		3.043			37				
		2.985			40				
		2.72			92.5				Anders & Jensen
Debye Temperature		388		°K	34				Koch & Scarbrough
		384			37				
		371			40				
		416			92.5				Anders & Jensen
Magnetic Susceptibility	χ_g	6.3		10^{-7}emu	37				Bucher et al. C
Microhardness		965		kg/mm^2	34				Bucher et al. C
		1010			37				
		975			40				

NIOBIUM PLATINUM

PROPERTY	SYMBOL	VALUE	UNIT	AT.% Pt	CRYSTAL DATA	TEMP. (°K)	NOTES	REFERENCES
Lattice Parameters	a_o	5.153	Å	25				Geller et al.
	a_o	9.91						
	c_o	5.13		38	σ-phase			Bucher et al. C
	a_o	9.91		40				Greenfield & Beck
	c_o	5.12						
Density		11.50	g/cm^3	25				Zegler
		13.0		38				Bucher et al. C
Transition Temperature	T_c	10.9	°K	25			argon arc-melted	Zegler
		4.21		38				Bucher et al. F
Electronic Specific Heat		3.4	mJ/mol-°K^2	38				Bucher et al. F
Debye Temperature		336	°K	38				Bucher et al. F

BIBLIOGRAPHY

AIR FORCE SYSTEMS COMMAND. RES. AND TECHNOL. DIV. Emerging Aerospace Materials. Report No. AFML TR-65-114. Apr. 1965. 109 p.

ALEKSEEVSKII, N.E. and N.N. MIKHAILOV. Superconductivity of Some Binary and Ternary Alloys. SOVIET PHYS. JETP, v. 16, no. 6, June 1963. p. 1493-1495.

AMES, S.L. and A.D. McQUILLAN. The Resistivity-Temperature-Concentration Relationships in the System Niobium-Titanium. ACTA METAL., v. 2, no. 1, Nov. 1954. p. 831-836.

ANDERSSON, L.H. and R. KIESSLING. Investigations on the Binary Systems of Boron with Chromium, Columbium, Nickel, and Thorium, Including a Discussion of the Phase "TiB" in the Titanium-Boron System. ACTA CHEMICA SCANDINAVICA, v. 4, 1950. p. 160-164.

ANDRES, K. and M.A. JENSEN. Superconductivity, Susceptibility, and Specific Heat in the Noble Transition Elements and Alloys. I. Experimental Results. PHYS. REV., v. 165, no. 2, Jan. 10, 1968. p. 533-544.

ARMSTRONG, G.T. The Low Temperature Heat Capacity of Columbium Nitride. AMERICAN CHEM. SOC., J., v. 71, no. 11, Nov. 17, 1949. p. 3583-3587.

ARON, P.R. and H.C. HITCHCOCK. Anomalous Critical Currents in Niobium-25 Atomic % Zirconium Wire. In: SUPERCON-DUCTORS. Edited by M. Tanenbaum and W.V. Wright. New York, Interscience, Feb. 18, 1962. p. 115-121. B

ARON, P.R. and H.C. HITCHCOCK. Critical Currents of Superconducting Niobium-25 Atomic % Zirconium in High Magnetic Fields. J. OF APPLIED PHYS., v. 33, no. 7, July 1962. p. 2242-2244. A

ASCHERMANN, G. et al. Superconducting Compounds with Very High Transition Temperatures (Niobium Hydride and Niobium Nitride) (In Ger.). PHYSIKALISCHE ZEIT., v. 42, no. 21/22, Nov. 20, 1941. p. 349-360.

AVGUSTINIK, A.I. et al. Conductivity Mechanism in Deformed Niobium Monocarbide (In Russ.). AKAD. NAUK SSSR. IZV. NEORGAN. MAT., v. 3, no. 2, 1967. p. 286-290.

BACHNER, F.J. and H.C. GATOS. Superconductivity Degradation in Beta-Tungsten Structure Compounds-Niobium Stannide and Niobium Aluminide. AIME METALLURGICAL SOC., TRANS., v. 236, no. 9, Sept. 1966. p. 1261-1266.

BACHNER, F.J. et al. Superconducting Transition Temperature and Electronic Structure in the Pseudobinaries Niobium Aluminum Stannide and Niobium Tin Antimonide. J. OF PHYS. AND CHEM. OF SOLIDS, v. 28, no. 5, May 1967. p. 889-895.

BAKER, C. and J. SUTTON. Correlation of Superconducting and Metallurgical Properties of a Titanium-20 Atomic % Niobium Alloy. PHIL. MAG., v. 19, no. 162, June 1969. p. 1223-1255.

BANUS, M.D. et al. Nb_3In: A Beta-Tungsten Structure Superconducting Compound. J. OF PHYS. AND CHEM. OF SOLIDS, v. 23, July 1962. p. 971-973.

BARNES, L.J. Tunneling at Point Contacts Between Superconductors. PHYS. REV., v. 184, no. 2, Aug. 10, 1969. p. 434-446.

BEERNTSEN, D.J. et al. Anisotropic Superconducting Properties of $NbSe_2$. IEEE TRANS. ON AEROSPACE, v. 2, no. 2, Apr. 1964. p. 816-821.

BENDER, D. et al. Structure and Electronic Properties of Niobium-Ruthenium Alloys. PHYS. KONDENS. MATERIE, v. 1, no. 3, 1963. p. 225-231.

BERLINCOURT, T.G. Pulsed Magnetic Field Studies of Superconducting Transition Metal Alloys at High and Low Current Densities. INTERNAT. CONF. ON LOW TEMP. PHYS., PROC., 8TH. Ed. by Davies, R.O. Sept. 16-22, 1962. Pub. Butterworth, Inc., Washington, D.C. 1963. p. 338-341.

BERLINCOURT, T.G. and R.R. HAKE. Superconductivity at High Magnetic Fields. PHYS. REV., v. 131, no. 1, July 1, 1963. p. 140-157.

BERNASSON, M. et al. Magnetic Properties of 75% Niobium-25% Metal. SOLID STATE COMM., v. 8, no. 11, June 1970. p. 837-841.

BINDARI, A.El. and M.M. LITVAK. The Upper Critical Field of Niobium-Zirconium and Niobium-Titanium Alloys. Rept. No. BSD-TDR-63-32. Contract No. AF 04(694)-33. Jan. 1963. 7 p. AD 299781

BLAUGHER, R.D. and J.K. HULM. Superconductivity in the Sigma and Alpha-Manganese Structures. J. OF PHYS. AND CHEM. OF SOLIDS, v. 19, no. 1/2, 1961. p. 134-138.

BLAUGHER, R.D. et al. Factors Affecting Superconductivity in Alloys with Unfilled d-Bands. In: INTERNAT. CONF. ON LOW TEMP. PHYS., PROC., 8TH. Sept. 1962. p. 147-148.

BOOM, R.W. and L.D. ROBERTS. Study of the Transition of Small Niobium-Zirconium Superconducting Solenoids to the Normal State. J. OF APPLIED PHYS., v. 34, no. 8, Aug. 1963. p. 2422-2425.

BORODICH, V.D. et al. The Critical Currents of Niobium-Zirconium Alloys in an External Magnetic Field. SOVIET PHYS. JETP, v. 17, no. 1, July 1963. p. 76-79.

BORUKHOVICH, A.S. et al. Magnetic Susceptibility of Niobium and Tantalum Monocarbides at Low Temperatures. SOVIET PHYS. SOLID STATE, v. 11, no. 3, Sept. 1969. p. 681-682.

BOSOMWORTH, D.R. and G.W. CULLEN. Energy Gap of Superconducting Niobium Stannide. PHYS. REV., v. 160, no. 2, Aug. 10, 1967. p. 346-347.

BRAENDLI, G. and R. GRIESSEN. Magnetostriction Due to Surface Currents in Type-II Superconductors. PHYS. REV. LETTERS, v. 22, no. 11, Mar. 17, 1969. p. 534-537.

BRANDT, N.B. and N.I. GINZBURG. The Effect of High Pressure on the Superconducting Transition Temperature of the Alloys 90% Molybdenum-10% Rhenium and 75% Niobium-25% Molybdenum. SOVIET PHYS. JETP, v. 24, no. 1, Jan. 1967. p. 40-41.

BRAUER, G. and J. JANDER. The Niobium Nitrides (In Ger.). Z. FUER ANORG. UND ALLGEM. CHEM., v. 270, 1952. p. 160-178.

BRENTON, R.F. et al. Elastic Properties and Thermal Expansion of Niobium Monocarbide to High Temperatures. J. OF LESS COMMON METALS, v. 19, no. 3, Nov. 1969. p. 273-278.

BREWER, L. et al. A Study of the Refractory Borides. AMERICAN CERAM. SOC., J., v. 34, no. 6, June 1951. p. 173-179.

BRIXNER, L.H. Preparation and Properties of the Single Crystalline AB_2-Type Selenides and Tellurides of Niobium, Tantalum, Molybdenum and Tungsten. J. OF INORGANIC NUCLEAR CHEMISTRY, v. 24, 1962. p. 257-263.

BROWN, A.R.G. et al. The Titanium-Niobium System. NATURE, v. 201, no. 4922, Feb. 1964. p. 914-915.

BROWN, H.L. et al. Elastic Properties of Some Polycrystalline Transition-Metal Monocarbides. J. OF CHEM. PHYS., v. 45, no. 2, July 15, 1966. p. 547-549.

BUCHER, E. and J. MUELLER. Superconductivity in Hexagonal Titanium-Vanadium and Titanium-Niobium Alloys (In Ger.). HELV. PHYS. ACTA, v. 34, no. 4, July 1, 1961. p. 410-413.

BUCHER, E. et al. Electronic Specific Heat and Superconductivity of Niobium-Ruthenium Alloys. INTERNAT. CONF. ON LOW TEMP. PHYS., PROC., 8TH. Ed. by Davies, R.O. Sept. 16-22, 1962. Pub. Butterworth, Inc., Washington, D.C., 1963. p. 151-152. A

BUCHER, E. et al. Superconductivity and Electronic Properties of Binary Complex Phases of the Transition Metals. INTERNAT. CONF. ON LOW TEMP. PHYS., PROC., 8TH. Ed. by Davies, R.O. Sept. 16-22, 1962. Pub. Butterworth, Inc., Washington, D.C., 1963. p. 153-154. B

BUCHER, E. et al. Superconductivity and Paramagnetism in Complex Phases of the Transition Metals (In Ger.). HELV. PHYS. ACTA, v. 34, no. 8, Dec. 31, 1961. p. 843-858. C

BUCHER, E. et al. Superconductivity and Electronic Properties of Transition Metal Alloys. REV. OF MODERN PHYS., v. 36, no. 1, Pt. 1, Jan. 1964. p. 146-149. D

BUCHER, E. et al. A Phase Transition and Its Influence on Superconductivity in the (Niobium, Vanadium)-Gold A-15-Type Structure. PHYS. LETTERS, v. 8, no. 1, Jan. 1, 1964. p. 27-28. E

BUCHER, E. et al. Specific Heat and Magnetic Susceptibility in Superconductive Binary Complex Phases of Transition Metals (In Ger.). PHYS. KONDENS. MATERIE, v. 2, no. 3, 1964. p. 210-240. F

BUEHLER, E. et al. Preparation and Properties of Cored Wire Containing Niobium Stannide and V_3Ga. In: METALLURGY OF ADVANCED ELECTRONIC MATERIALS, PROCEEDINGS., Philadelphia, Aug. 27-29, 1962. Interscience Publishers, 1963. p. 105-119.

BYCHKOV, Yu.F. et al. Influence of the Structural State on the Superconducting Properties of Zirconium Alloys Containing 20 to 25% Niobium. SOVIET PHYS. JETP, v. 21, n o. 3, Sept. 1965. p. 543-547.

CALVERLEY, A. and A.C. ROSE INNES. Trapped Flux in Superconducting Mixed Crystals. ROYAL SOC. OF LONDON, PROC., A, v. 255, no. 1281, Apr. 1960. p. 267-276.

CAPPELLETTI, R.L. et al. Far-Infrared Absorption in Superconducting Niobium Alloys. PHYS. REV., v. 158, no. 2, June 10, 1967. p. 340-345.

CARPENTER, J.H. and A.W. SEARCY. Preparation, Identification, and Chemical Properties of the Niobium Germanides. AMERICAN CHEM. SOC., J., v. 78, no. 10, May 20, 1956. p. 2079-2081.

CHANDRASEKHAR, B.S. et al. The High Field Superconductivity of Niobium-Zirconium Alloys. INTERNAT. CONF. ON LOW TEMP. PHYS., PROC., 8TH. Ed. by Davies, R.O. Sept. 16-22, 1962. Pub. Butterworth, Inc., Washington, D.C., 1963. p. 345-347.

CHARLESWORTH, J.P. et al. Experimental Work on the Niobium-Tin Constitution Diagram and Related Studies. J. OF MATERIALS SCI., v. 5, no. 7, July 1970. p. 580-603.

CHERRY, W.H. Surge-Magnetic-Field and Pulse-Current Effects in Niobium Stannide. RCA REVIEW, v. 25, no. 3, Sept. 1964. p. 510-532.

CHERRY, W.H. et al. Superconductivity in Metals and Alloys. Technical Documentary Rept. No. ASD-TDR-62-1111. Contract No. AF 33-657-7733. Feb. 1963. 60 p. A

CHERRY, W.H. et al. Superconductivity in Metals and Alloys. Technical Rept. 60-919. Contract No. AF 33(616)-6405. June 1961. 53 p. B

CODY, G.D. and R.W. COHEN. Thermal Conductivity of Nb$_3$Sn. REV. OF MODERN PHYS., v. 36, no. 1, Pt. 1, Jan. 1964. p. 121-123.

CODY, G.D. and G.W. CULLEN. Critical Currents and Lorentz-Force Model in Niobium Stannide. RCA REVIEW, v. 25, no. 3, Sept. 1964. p. 466-478.

CODY, G.D. The Superconducting Penetration Depth of Niobium Stannide. RCA REVIEW, v. 25, no. 3, Sept. 1964. p. 414-432.

CODY, G.D. et al. Phenomenon of Superconductivity. Technical Rept. AFML-TR-65-169. Contract No. AF 33-657-11208. June 1965. 242 p. AD 465 438.

COFFEY, H.T. et al. Effect of Low-Temperature Deuteron Irradiation on Some Type-II Superconductors. PHYS. REV., v. 155, no. 2, Mar. 10, 1967. p. 355-363.

COHEN, R.W. et al. The Superconducting Energy Gap of Niobium Stannide. RCA REVIEW, v. 25, no. 3, Sept. 1964. p. 433-452.

COMPTON, V.B. et al. Superconductivity of Technetium Alloys and Compounds. PHYS. REV., v. 123, no. 5, Sept. 1, 1961. p. 1567-1568.

COOK, D.B. et al. Superconductivity of Columbium Nitride. PHYS. REV., v. 79, no. 6, Sept. 15, 1950. p. 1021.

COOPER, J.L. Transition Temperature of Niobium Stannide. RCA REVIEW, v. 25, no. 3, Sept. 1964. p. 405-413.

CORENZWIT, E. Superconductivity of Nb$_3$Al. PHYS. AND CHEM. OF SOLIDS, v. 9, no. 1, 1959. p. 93.

CORSAN, J.M. and A.J. COOK. Electronic Specific Heat and Superconducting Properties of Niobium-Tantalum Alloys. PHYS. LETTERS, v. 28A, no. 7, Jan. 13, 1969. p. 500-501.

COURTNEY, T.H. et al. The Influence of Point Defects on Some Superconducting Properties of Nb$_3$Sn. AIME METALL. SOC., TRANS., v. 233, no. 1, Jan. 1965. p. 212-218.

DARNELL, J.R. and L.F. YNTEMA. The Element Columbium and its Compounds. In: SYMPOSIUM ON COLUMBIUM - NIOBIUM. Washington, D.C. Technology of Columbium - Niobium papers presented at the Symposium on Columbium - Niobium. Ed. by Gosner, B. and E. Sherwood. N.Y., Wiley, 1958. p. 1-9.

DAVISSON, J.W. and J. PASTERNAK. Status Report on Superconductivity. NRL Memorandum Rept. 1089, Quarterly Status Rept. No. 5. Aug. 1960. 61 p.

DEARDORFF, D.K. et al. New Tetragonal Compounds Nb$_3$Si and Ta$_3$Si. J. OF LESS COMMON METALS, v. 18, no. 1, May 1969. p. 11-26.

DeSORBO, W. The Peak Effect in Substitutional and Interstitial Solid Solutions of High-Field Superconductors. REV. OF MODERN PHYS., v. 36, no. 1, Pt. 1, Jan. 1964. p. 90-94.

DeSORBO, W. Size Factor and Superconducting Properties of Some Transition Metal Solutions. PHYS. REV., v. 130, no. 6, June 15, 1963. p. 2177-2187.

DeSORBO, W. Effect of Dissolved Gases on Some Superconducting Properties of Niobium. PHYS. REV., v. 132, no. 1, Oct. 1, 1963. p. 107-121.

DIETRICH, I. et al. Studies on Wires of Superconducting Alloys in the Niobium-Titanium and Niobium-Zirconium Systems (In Ger.). Z. FUER METALLK., v. 53, 1962. p. 721-728.

DONNAY, J.D.H. Crystal Data. Determinative Tables. 2nd Edition. American Crystallographic Association, 1963.

DRUZHININA, I.P. et al. Thermoelectric Properties of the Alloys of Niobium with Titanium and Hafnium. MEASUREMENT TECHNIQUES, no. 10, Oct. 1967. p. 1215-1216.

DUBECK, L. and K.S.L. SETTY. Thermal Conductivity of Niobium-Titanium. PHYS. LETTERS, v. 27A, no. 6, Aug. 12, 1968. p. 334-335.

DUBROVSKAYA, L.B. et al. The Magnetic Susceptibility of Solid Solutions of Zirconium and Niobium Monocarbides. SOVIET PHYS. SOLID STATE, v. 11, no. 10, Apr. 1970. p. 2451-2453.

DUWEZ, P. The Allotropic Transformation of Hafnium. J. OF APPLIED PHYS., v. 22, no. 9, Sept. 1951. p. 1174-1175.

DWIGHT, A.E. Alloying Behavior of Columbium. In: COLUMBIUM METALLURGY. Edited by Douglass, D.L. and F.W. Kunz. New York, Interscience Publishers, 1961. p. 383-406.

DYUBUYA, B.Ch. and O.K. KULTASHEV. Work Function of Alloys W-Hf, Ta-Hf, Nb-Hf, Re-Hf, Re-Zr and W-Re. PHYS. OF METALS AND METALL., v. 21, no. 3, 1966. p. 76-82.

EDGECUMBE, J. et al. Preparation and Properties of Thin-Film Hard Superconductors. J. OF APPLIED PHYS., v. 35, no. 7, July 1964. p. 2198-2202.

EDWARDS, J.W. et al. High Temperature Structure and Thermal Expansion of Some Metals as Determined by X-Ray Diffraction Data. I. Platinum, Tantalum, Niobium and Molybdenum. J. OF APPLIED PHYS., v. 22, no. 4, Apr. 1951. p. 424-428.

ELLIOTT, R.P. Columbium-Carbon System. AMERICAN SOC. METALS, TRANS., v. 53, 1961. p. 13-28.

ELLIS, T.G. and H.A. WILHELM. Phase Equilibria and Crystallography for the Niobium-Tin System. J. OF LESS COMMON METALS, v. 7, no. 1, July 1964. p. 67-83.

ENSTROM, R.E. Superconducting Properties of 55% Niobium-45% Tin and of Multiphase Niobium-Tin Alloys. J. OF APPLIED PHYS., v. 37, no. 13, Dec. 1966. p. 4880-4882.

ENSTROM, R.E. et al. Superconductivity of Two-Phase Niobium-Niobium Tin Sintered Compacts. APPLIED PHYS. LETTERS, v. 3, no. 5, Sept. 1, 1963. p. 81-82. A

ENSTROM, R. et al. Current and Field Dependence of Superconductivity on Microstructure in the Niobium-Tin System. In: METALLURGY OF ADVANCED ELECTRONIC MATERIALS, PROCEEDINGS. Philadelphia, Aug. 27-29, 1962. Interscience Publishers, 1963. p. 121-150. B

EVANS, D.J. and R.A. ERICKSON. Electrical Resistivity of Niobium-Zirconium Alloys Below 273.2°K. J. OF APPLIED PHYS., v. 36, no. 11, Nov. 1965. p. 3517-3520.

FIETZ, W.A. and W.W. WEBB. Magnetic Properties of Some Type-II Alloy Superconductors near the Upper Critical Field. PHYS. REV., v. 161, no. 2, Sept. 1967. p. 423-433 A

FIETZ, W.A. and W.W. WEBB. Hysteresis in Superconducting Alloys-Temperature and Field Dependence of Dislocation Pinning in Niobium Alloys. PHYS. REV., v. 178, no. 2, Feb. 10, 1969. p. 657-667. B

FOX, D.K. and W.J. REICHENECKER. New Data on Superconducting Alloys. MATERIALS IN DESIGN ENGINEERING, v. 57, no. 4, Apr. 1963. p. 92-93.

FRENCH, R.A. et al. Almost Ideal Behaviour in Some Type II Superconducting Alloys. CRYOGENICS, v. 7, no. 2, Apr. 1967. p. 83-88.

FURUSETH, S. and A. KJEKSHUS. The Crystal Structures of Niobium Arsenide and Niobium Antimonide. ACTA CRYST., v. 18, 1965. p. 320-324. A

FURUSETH, S. and A. KJEKSHUS. On the Arsenides and Antimonides of Niobium. ACTA CHEM. SCANDINAVICA, no. 18, 1964. p. 1180-1195. B

GAMBINO, J.R. Thermoelectric Properties of Refractory Materials. In: THERMOELECTRIC MATERIALS AND DEVICES. Edited by Cadoff, I. and E. Miller, N.Y., Reinhold, 1960. p. 163-172.

GANGULY, B.N. and K.P. SINHA. Pressure Effects on the Transition Temperature of Superconductors. INDIAN J. OF PURE AND APPL. PHYS., v. 4, no. 2, 1966. p. 49-56.

GAULE, G.K. et al. Superconductivity of the Molybdenum Borides and Related Materials. INTERNAT. CONF. ON LOW TEMP. PHYS., PROC., 8TH. Ed. by Davies, R.O. Sept. 16-22, 1962. Pub. Butterworth, Inc., Washington, D.C. 1963. p. 162-165.

GAUSTER, W.F. et al. Magnetic Flux Flow and Superconductor Stabilization. Rept. No. ORNL-TM-2233. May 17, 1968. 43 p. N68-33627.

GEBALLE, T.H. et al. High Temperature SP-Band Superconductors. PHYSICS, v. 2, no. 6, June 1966. p. 293-310.

GEBBHARDT, E. and R. ROTHENBACHER. Studies in the Niobium-Oxygen System (In Ger.). Z. FUER METALLK., v. 54, no. 8, Aug. 1963. p. 443-448.

GELLER, S. et al. Some New Intermetallic Compounds with the "Beta-Wolfram" Structure. AMERICAN CHEM. SOC., J., v. 77, n o. 5, Mar. 20, 1955. p. 1502-1504.

GIORGI, A.L. et al. Investigation of Ta$_2$C, Nb$_2$C, and V$_2$C for Superconductivity. PHYS. REV., v. 129, no. 4, Feb. 15, 1963. p. 1524-1525.   A

GIORGI, A.L. et al. Effect of Composition on the Superconducting Transition Temperature of Tantalum Carbide and Niobium Carbide. PHYS. REV., v. 125, no. 3, Feb. 1, 1962. p. 837-838.   B

GIORGI, A.L. et al. Anomalous Superconducting Properties of Refractory Carbides and Nitrides of Group IVa and Va Elements. Rept. No. LA-DC-8022. 8 p. 1966. N68-12413.   C

GITTLEMAN, J.I. et al. Phenomenon of Superconductivity (Phase II). Technical Rept. AFML-TR-67-138, Mar. 15, 1966-Mar. 15, 1967. Contract No. AF 33(615)-3936. June 1967. 105 p.

GLADYSHEVSKII, E.I. et al. Crystal Structures of Some Intermetallic Compounds. SOVIET PHYS. CRYST., v. 6, no. 2, Sept./Oct. 1961. p. 207-208.

GOLDSCHMIDT, H.J. and J.A. BRAND. The Constitution of the Chromium-Niobium-Molybdenum System. J. OF LESS COMMON METALS, v. 3, no. 1, Feb. 1961. p. 44-61.

GOLIKOVA, O.A. et al. Electrical Properties of Carbides of Group IV Transition Metals. SOVIET PHYS. SEMICON-DUCTORS, v. 3, no. 4, Oct. 1969. p. 429-433.   A

GOLIKOVA, O.A. et al. Electrical Conductivity and Thermo-EMF of Niobium Carbide Between 20 and 2000°C. HIGH TEMP., v. 5, no. 6, Nov./Dec. 1967. p. 894-896. B

GREENFIELD, P. and P.A. BECK. Intermediate Phases in Binary Systems of Certain Transition Elements. AIME METALL., SOC., TRANS., J. OF METALS, v. 8, no. 2, Feb. 1956. p. 265-276.

GUTS, Z.A. et al. Investigation of Superconducting Alloys in the Niobium-Zirconium System. SOVIET PHYS. SOLID STATE, v. 5, no. 1, July 1963. p. 262-264. A

GUTS, Z.A. et al. Superconducting Niobium-Gallium Alloys. SOVIET PHYS. TECH. PHYS., v. 10, no. 9, Mar. 1966. p. 1295-1296.   B

HAKE, R.R. Mixed-State Paramagnetism in High-Field Type-II Superconductors. PHYS. REV. LETTERS, v. 15, no. 22, Nov. 29, 1965. p. 865-868.   A

HAKE, R.R. Upper-Critical-Field Limits for Bulk Type-II Superconductors. APPLIED PHYS. LETTERS, v. 10, no. 6, Mar. 15, 1967. p. 189-192.   B

HAKE, R.R. et al. High-Field Superconducting Characteristics of Some Ductile Transition Metal Alloys. In: SUPERCONDUCTORS. Edited by M. Tanenbaum and W.V. Wright. New York, Interscience, Feb. 18, 1962. p. 53-60. A

HAKE, R.R. et al. High-Field Superconducting Characteristics of Some Ductile Transition Metal Alloys. In: SUPERCONDUCTORS. Edited by M. Tanenbaum and W.V. Wright. New York, Interscience, 1962. p. 53-60. A

HAKE, R.R. et al. Giant Anisotropy in the High Field Critical Currents of Cold Rolled Transition Metal Alloy Superconductors. INTERNAT. CONF. ON LOW TEMP. PHYS., PROC., 8TH. Ed. by Davies, R.O. Sept. 16-22, 1962. Pub. Butterworth, Inc., Washington, D.C. 1963. p. 342-344. B

HANAK, J.J. et al. Preparation and Properties of Vapor-Deposited Niobium Stannide. RCA REVIEW, v. 25, no. 3, Sept. 1964. p. 342-365.   A

HANAK, J.J. et al.   Some Physical Properties of Deposited Nb$_3$Sn. In: HIGH MAGNETIC FIELDS. Edited by H. Kolm et al. PROC. OF THE INTERNAT. CONF. ON HIGH MAGNETIC FIELDS, Held at MIT Nov. 1-4, 1961. MIT and Wiley, c1962. p. 592-596. b

HANSEN, M. Systems Titanium-Molybdenum and Titanium-Columbium. AIME METALL. SOC., TRANS., J. OF METALS, v. 3, no. 10, Oct. 1951. p. 881-888.

HANSEN, M. and K. ANDERKO. Constitution of Binary Alloys. Second Edition. McGraw Hill, New York, 1958.

HARDY, G.F. and J.K. HULM. The Superconductivity of Some Transition Metal Compounds. PHYS. REV., v. 93, no. 5, Mar. 1, 1954. p. 1004-1016.

HARDY, T.C. and W. MILLER. Excess Low Temperature Specific Heat Due to Iron in Molybdenum-Niobium. SOLID STATE COMM., v. 7, no. 17, Sept. 1969. p. 1229-1234.

HART, H.R., JR. et al. Superconducting Critical Current of Niobium Stannide in Pulsed Magnetic Fields. In: HIGH MAGNETIC FIELDS. Edited by H. Kolm et al. PROC. OF THE INTERNAT. CONF. ON HIGH MAGNETIC FIELDS, Held at MIT Nov. 1-4, 1961. MIT and Wiley, c1962. Chapter 70, p. 584-588.

HATT, B.A. and V.G. RIVLIN. Phase Transformations in Superconducting Ti-Nb Alloys. BRITISH J. OF APPL. PHYS., (J. OF PHYS., D), v. 1, Ser. 2, no. 9, Sept. 1968. p. 1145-1149.

HEATON, J.W. and A.C. ROSE-INNES. Current Capacity of a Superconductor of the Second Kind. APPLIED PHYS. LETTERS, v. 2, no. 10, May 15, 1963. p. 196-197.

HECHLER, K. et al. Critical Data of Niobium Nitride in Transverse Magnetic Fields. ZEIT. FUER PHYSIK, v. 205, 1967. p. 400-408.

HECHLER, K. and E. SAUR. Preparation and Superconducting Properties of Pure as Well as Zirconium and Titanium Alloyed Niobium Nitride (In Ger.). ZEIT. FUER PHYSIK, v. 205, 1967. p. 393-399.

HECHT, R. Lower Critical Field of Niobium Stannide. RCA REVIEW, v. 25, no. 3, Sept. 1964. p. 453-465.

HEIN, R.A. et al. Superconducting Behavior of A15 Compounds. SOLID STATE COMM., v. 7, no. 3, Feb. 1969. p. 381-384. A

HEIN, R.A. et al. Superconductivity of the Niobium Molybdenum Alloys at Temperatures Below 0.25°K. AMERICAN PHYS. SOC., BULL., v. 7, no. 1, Pt. 1, Jan. 24, 1962. p. 322. B

HEIN, R.A. et al. Superconductivity in the Niobium-Molybdenum System. REV. OF MODERN PHYS., v. 36, no. 1, Pt. 1, Jan. 1964. p. 149-152. C

HEINIGER, F. et al. Low Temperature Specific Heat of Transition Metals and Alloys. PHYS. KONDENS. MATERIE, v. 5, no. 4, 1966. p. 243-284.

HEINIGER, F. and J. MULLER. Bulk Superconductivity in Dilute Hexagonal Titanium Alloys. PHYS. REV., v. 134, Ser. 2, no. 6A, June 15, 1964. p. A1407-A1409.

HOFFSTEIN, V. and R.W. COHEN. The Anisotropic Superconducting Energy Gap of Niobium Stannide. PHYS. LETTERS, v. 29A, no. 10, Aug. 11, 1969. p. 603-604.

HORN, G. and E. SAUR. Preparation and Superconductive Properties in Niobium Nitride as Well as Niobium Nitride with Titanium, Zirconium, and Tantalum Additions (In Ger.). ZEIT. FUER PHYSIK, v. 210, no. 1, 1968. p. 70-79.

HORN, F.H. and W.T. ZIEGLER. Superconductivity and Structure of Hydrides and Nitrides of Tantalum and Columbium. AMERICAN CHEM. SOC., J., v. 69, no. 11, Dec. 4, 1947. p. 2762-2769.

HOUSKA, C.R. Thermal Expansion of Certain Group IV and Group V Carbides at High Temperatures. AMERICAN CERAM. SOC., J., v. 47, no. 6, June 1964. p. 310-311.

HUDSON, W.R. Effect of Tensile Stress on Current-Carrying Capacity of Commercial Superconductors. Rept. No. NASA TN D-3745. Nov. 1966. 12 p.

HULLIGER, F. New Representatives of the $NbAs_2$ and $ZrAs_2$ Structures. NATURE, v. 204, no. 4960, Nov. 21, 1964. p. 775.

HULM, J.K. and B.T. MATTHIAS. New Superconducting Borides and Nitrides. PHYS. REV., v. 82, no. 2, Apr. 15, 1951. p. 273-274.

HULM, J.K. and R.D. BLAUGHER. Superconducting Solid Solution Alloys of the Transition Elements. PHYS. REV., v. 123, no. 5, Sept. 1, 1961. p. 1569-1580.

HURLEY, G.F. and J.H. BROPHY. A Constitution Diagram for the Niobium-Ruthenium System Above 1100°C. J. OF LESS COMMON METALS, v. 7, no. 4, Oct. 1964. p. 267-277.

IKUSHIMA, A. and T. MIZUSAKI. Superconductivity in Niobium and Niobium-Tantalum Alloys. J. OF PHYS. AND CHEM. OF SOLIDS, v. 30, no. 4, Apr. 1969. p. 873-879.

ISHIKAWA, M. and L.E. TOTH. Electronic Specific Heats and superconductivity in the Group-V Transition Metals. PHYS. REV., B, (SOLID STATE PHYS.), v. 3, no. 6, Mar. 1971. p. 1856-1861.

ITSKEVICH, E.S. et al. Effect of Pressure on the Superconducting Transition Temperature of the Alloys Nb_3Sn and Niobium-Zirconium. SOVIET PHYS. JETP, v. 18, no. 4, Apr. 1964. p. 949-950.

JELLINEK, F. et al. Molybdenum & Niobium Sulphides. NATURE, v. 185, no. 4710, Feb. 1960. p. 376-377.

JONES, C.K. et al. Upper Critical Field of Solid Solution Alloys of the Transition Elements. REV. OF MODERN PHYS., v. 36, no. 1, Pt. 1, Jan. 1964. p. 74-76.

KELLER, K.R. and J.J. HANAK. Ultrasonic Measurements in Single-Crystal Niobium Stannide. PHYS. REV., v. 154, no. 3, Feb. 15, 1967. p. 628-632.

KENDALL, E.G. and J.D. McCLELLAND. Updating the Refractory Materials. MACHINE DESIGN, v. 36, no. 25, Oct. 22, 1964. p. 208-218.

KIEFFER, R. et al.   Tungsten Alloys of High Melting Point.   J. OF LESS COMMON METALS, v. 1, 1959.   p. 19-33. A

KIEFFER, R. et al.   Formulation of the Vanadium-Silicon and Niobium-Silicon Systems (In Ger.).   Z. FUER METALLK., v. 47, no. 4, 1956.   p. 247-253.   B

KILLPATRICK, D.H.   Pressure-Temperature Phase Diagrams for Nb_3In and Nb_3Bi.   J. OF PHYS. AND CHEM. OF SOLIDS, v. 25, no. 11, Nov. 1964.   p. 1213-1216.

KIM, Y.B. et al.   Flux-Flow Resistance in Type-II Superconductors.   PHYS. REV., v. 139, no. 4A, Aug. 16, 1965. p. A1163-A1172.

KIMURA, Y. et al.   The Normal State Specific Heat of Niobium-Tantalum Alloys.   PHYS. LETTERS, v. 29A, no. 5, May 19, 1969.   p. 284-285.

KING, A.H. et al.   Effect of the Omega Transformation on Critical Currents in a Zirconium-Rich Niobium Alloy. CRYOGENICS, v. 5, no. 4, Aug. 1965.   p. 230-231.

KNAPTON, A.G.   An X-Ray Survey of Certain Transition-Metal Systems for Sigma Phases.   INST.OF METALS, J., v. 87, pt. 1, Sept. 1958.   p. 28-32.   A

KNAPTON, A.G.   Niobium and Tantalum Alloys.   J. OF LESS COMMON METALS, v. 2, no. 1, Feb. 1960.   p. 113-124. B

KNEIP, G.D., JR. et al.   Increased Critical Currents in Niobium-Zirconium Superconductors from Precipitation-Induced Defects.   In: HIGH MAGNETIC FIELDS.   Edited by H. Kolm et al.   PROC. OF THE INTERNAT. CONF. ON HIGH MAG-NETIC FIELDS, Held at MIT Nov. 1-4, 1961.   MIT and Wiley, c1962.   Chapter 74, p. 603-608.

KOCH, C.C. and G.R. LOVE.   Superconductivity in Niobium Containing Ferromagnetic Gadolinium or Paramagnetic Yttrium Dispersions.   J. OF APPLIED PHYS., v. 40, no. 9, Aug. 1969.   p. 3582-3587.

KOCH, C.C. and J.O. SCARBROUGH.   Superconductivity in Molybdenum-Rhenium and Niobium-Iridium Sigma Phases. PHYS. REV., B, v. 3, no. 3, Feb. 1971.   p. 742.

KOCH, D. et al.   Preparation and Superconductivity Properties of Niobium-Tin Harnesses (In Ger.).   ZEIT. FUER PHYSIK, v. 180, 1964.   p. 476-482.

KOMENOU, K. et al.   Energy Gap Measurement of Niobium Nitride.   PHYS. LETTERS, v. 28A, no. 5, Dec. 16, 1968. p. 335-336.

KOTLYAR, A.A. and Ts.V. VOSKOBOINIK.   Determination of the Spectral Coefficients of the Radiation of Tantalum-Niobium Alloys.   HIGH TEMP., v. 6, no. 5, Sept./Oct. 1968.   p. 794-796.

KRIVKO, N.I.   Localized Magnetic Moments in Niobium-Cobalt and Zirconium-Cobalt Alloys.   SOVIET PHYS. SOLID STATE, v. 11, no. 2, Aug. 1969.   p. 334-336.

KROEGER, D.M.   A Peak Effect in Critical Current with Respect to Temperature and Field in a Type-II Superconductor. SOLID STATE COMM., v. 7, no. 11, June 1969.   p. 843-847.

KULVARSKAYA, B.S. et al.   Thermionic Emission of Certain Refractory Materials and their Possible Application in Devices Filled with Complex Gaseous Medium.   RADIO ENG. AND ELECTRONIC PHYS., v. 13, no. 7, July 1968.   p. 1131-1134.   A

KULVARSKAYA, B.S. et al.   Thermionic Emission of Certain Refractory Materials and Possible Use as Cathodes in Gaseous Devices.   SOVIET PHYS. TECH. PHYS., v. 14, no. 1, July 1969.   p. 122-128.   B

KUNZ, W. and E. SAUR.   Superconductivity in the Systems Niobium-Tin, Niobium-Aluminum, Vanadium-Silicon and Vanadium-Gallium.   In: INT. CONF. ON LOW TEMP. PHYS., PROC., 9TH, Pt. A.   N.Y., Plenum Press, 1965.   p. 581-583.   B

KUNZ, W. and E. SAUR.   Superconducting of Fused Samples in the Systems Niobium-Tin, Niobium-Aluminum, Niobium-Galliu, Vanadium-Silicon and Vanadium-Gallium (In Ger.).   Z. FUER PHYSIK, v. 189, no. 4, 1966.   p. 401-416.   A

KUNZLER, J.E.   Superconductivity in High Magnetic Fields at High Current Densities.   REV. OF MODERN PHYS., v. 33, no. 4, Oct. 1961.   p. 501-509.

LAM, D.J. et al.   Nuclear Magnetic Resonance and Magnetic Susceptibilities of Vanadium-Niobium Alloys.   PHYS. REV., v. 156, no. 3, Apr. 15, 1967.   p. 735-739.

LAZAREV, B.G. et al.   Influence of Hydrostatic Compression on the Superconducting Transition Temperature of Nb_3Sn.   SOVIET PHYS. JETP, v. 16, no. 6, June 1963.   p. 1631-1632.

LeBLANC, M.A.R. and D.J. GRIFFITHS.   Hysteretic Phenomena in Type II Superconductors.   APPLIED PHYS. LETTERS, v. 9, no. 3, Aug. 1, 1966.   p. 131-134.

LEKSINA, I.Ye. et al.   Optical Properties of Superconductive Industrial Alloys of Niobium with Titanium.   PHYS. OF METALS AND METALL., v. 23, no. 3, 1967.   p. 116-120.

LEVESQUE, P. et al. The Constitution of Rhenium-Columbium Alloys. AMERICAN SOC. FOR METALS, TRANS., v. 53, 1961. p. 215-226.

LOWELL, J. et al. Thermally Induced Voltages in the Mixed State of Type II Superconductors. PHYS. LETTERS, v. 24A, no. 7, Mar. 27, 1967. p. 376-377.

LOWELL, J. Thermal Conductivity of Superconducting Tantalum-Niobium Alloys. PHYS. LETTERS, v. 22, no. 1, July 15, 1966. p. 11-12.

LVOV, S.N. et al. Some Regularities in the Electrical Properties of the Borides, Carbides, and Nitrides of the Transition Metals in Groups IV-VI of the Periodic Table. SOVIET PHYS. DOKL., v. 135, no. 3, Nov. 1960. p. 1334-1337.

MASUDA, Y. et al. Nuclear Magnetic Resonance and Specific Heat Measurements of Transition Metals and Alloys; Titanium-Vanadium-Iron and Zirconium-Niobium-Molybdenum Systems. PHYS. SOC. OF JAPAN, J., v. 22, no. 1, Jan. 1967. p. 238-247.

MATSKEVICH, T.L. and T.V. KRACHINO. Thermoelectron Emission of Some Refractory Compounds. SOVIET PHYS. TECH. PHYS., v. 7, no. 2, Aug. 1962. p. 156-158.

MATTHIAS, B.T. Empirical Relation Between Superconductivity and the Number of Valence Electrons per Atom. PHYS. REV., v. 97, no. 1, Jan. 1, 1955. p. 74-76.

MATTHIAS, B.T. and J.K. HULM. A Search for New Superconducting Compounds. PHYS. REV., v. 87, no. 5, Sept. 1, 1952. p. 799-806.

MATTHIAS, B.T. et al. Superconductivity. REV. OF MODERN PHYS., v. 35, no. 1, Jan. 1963. p. 1-22. A

MATTHIAS, B.T. et al. Superconductivity of Nb_3Ge. PHYS. REV., v. 139, no. 5A, Aug. 30, 1965. p. A1501-A1503. B

MATTHIAS, B.T. et al. Some New Superconducting Compounds. J. OF PHYS. AND CHEM. OF SOLIDS, v. 19, no. 1/2, 1961. p. 130-133. C

MAYKUTH, D.J. et al. The Thermoelectric Properties of Tantalum Alloys. AIME METALL. SOC., TRANS., v. 233, no. 6, 1965. p. 1196-1197.

McCONVILLE, T. and B. SERIN. Specific Heat of Type II Superconductors in a Magnetic Field. PHYS. REV. LETTERS, v. 13, no. 12, Sept. 21, 1964. p. 365-367.

MEYER, G. Penetration by a Weak Magnetic Field in Superconductors of 3 to 1 Niobium-Tin, Niobium-Aluminum and Vanadium Silicon (In Ger.). ZEIT. FUER PHYSIK, v. 189, no. 7, 1966. p. 199-206. [A]

MEYER, G. Experimental Determination of the Parameters of Superconductivity from Magnetization Measurements for Some beta-Tungsten Compounds (In Ger.). Z. FUER PHYSIK, v. 219, no. 4, 1969. p. 397-410. [B]

MEYER, G. and H. WIZGALL. Influence of Current and Magnetic Field on Superconducting Properties of Niobium-Tin, Vanadium-Gallium and Vanadium-Silicon Diffusion Layers (In Ger.). ZEIT. FUER PHYSIK, v. 183, no. 4, Mar. 8, 1965. p. 412-423.

MONTGOMERY, D.B. and W. SAMPSON. Measurements on Niobium-Tin Samples in 200-kG Continuous Fields. APPLIED PHYS. LETTERS, v. 6, no. 6, Mar. 15, 1965. p. 108-111.

MORIN, F.J. and J.P. MAITA. Specific Heats of Transition Metal Superconductors. PHYS. REV., v. 129, no. 3, Feb. 1, 1963. p. 1115-1120.

MÜLLER, C.B. and E.J. SAUR. Influence of Pressure on the Superconductivity of Some High-Field Superconductors. REV. OF MODERN PHYS., v. 36, no. 1, Jan. 1964. p. 103-105.

MYERS, G.E. and G.L. MONTET. Optical Properties of Single Crystals of Two Niobium Selenides. J. OF APPLIED PHYS., v. 41, no. 11, Oct. 1970. p. 4642-4649.

NESHPOR, V.S. and G.V. SAMSONOV. Electrical Thermoelectric and Galvanomagnetic Properties of Silicides and Transition Metals. SOVIET PHYS. DOKL., v. 5, no. 4, Jan.-Feb. 1961. p. 877-880.

NESHPOR, V.S. et al. Effect of Chemical Composition on the Electrothermophysical Properties of Zirconium and Niobium Carbide in the Region of Homogeneity (In Russ.). ZH. PRIKL. KHIM., v. 37, no. 11, 1964. p. 2375-2382. [A]

NESHPOR, V.S. et al. Study of the Chemical Properties of Homogeneous Monocarbides of Transition Metals in Groups IV and V in a Temperature Range as a Function of Specific Conductivity and Thermal EMF (In Russ.). AKAD. NAUK SSSR. IZV. NEORGAN. MAT., v. 2, no. 5, 1966. p. 855-863. [B]

NEUBAUER, H. Superconductive Transitions of Different Type II Compounds Under Hydrostatic Pressure (In Ger.). Z. FUER PHYSIK, v. 226, no. 3, 1969. p. 211-221.

NEVITT, M.V.   Atomic Size Effects in Cr_3O-Type Structure.   AIME METALL. SOC., TRANS., v. 212, June 1958. p. 350-355.

NIESSEN, A.K. and F.A. STAAS.   Hall Effect Measurements on Type II Superconductors.   PHYS. LETTERS, v. 15, no. 1, Mar. 1, 1965.   p. 26-28.

NOWOTNY, H. et al.   Structure of Some Germanides of Formula M_5Ge_3.   J. OF PHYS. CHEM., v. 60, no. 5, May 1956.   p. 677.

OAK RIDGE NAT. LAB., METALS AND CERAM. DIV.   Annual PR for period Ending June 30, 1964.   ORNL-3670.   Contract no. W-7405-eng-26.   Oct. 1964.

OGASAWARA, T. et al.   Magnetic Properties of Superconducting Niobium-Tantalum Alloys.   PHYS. SOC. OF JAPAN, J., v. 25, no. 5, Nov. 1968.   p. 1307-1323.

OLSEN, K.M. et al.   Preparation and Properties of Ultrafine Niobium-Zirconium Superconducting Wire.   In: SUPERCONDUCTORS.   Edited by M. Tanenbaum and W.V. Wright.   New York, Interscience, Feb. 18, 1962.   p. 123-128.

OTTO, G.   Superconductivity of Some Solid Solutions Between Compounds of A15 Structure on Niobium Base (In Ger.). ZEIT. FUER PHYSIK, v. 215, no. 4, 1968.   p. 323-334.

PADERNO, Yu.B. et al.   Nature of the Thermal Conductivity in the Carbides of Some Transition Metals of Groups IV and V.   SOVIET POWDER METALL. AND METAL CERAM., v. 7, no. 2, Feb. 1967.   p. 131-134.

PETERS, B.C. and A.A. HENDRICKSON.   Solid Solution Strengthening in Niobium-Tantalum and Niobium-Molybdenum Alloy Single Crystals.   METALLURGICAL TRANS., v. 1, no. 8, Aug. 1970.   p. 2271-2280.

PETROV, V.A. et al.   Total Hemispherical Emissive Power, Monochromatic (Lambda = 0.65 micron) Emissive Power, and Specific Electrical Resistivity of Zirconium and Niobium Carbides in the Temperature Range 1200-3500°K. HIGH TEMP., v. 5, no. 6, Nov./Dec. 1967.   p. 889-893.

PIPER, J.   Low Critical Currents in Superconducting Niobium Carbide.   APPLIED PHYS. LETTERS, v. 6, no. 9, May 1, 1965.   p. 183-184.   A

PIPER, J.   Electrical Properties of Some Transition-Metal Carbides and Nitrides.   Technical Rept. No. C-21. Contract No. DA-30-069-ORD-2787.   Apr. 1964.   34 p. AD 435 624   B

POPOV, I.A. and I.G. RODIONOVA.   The Molybdenum-Niobium-Zirconium System.   RUSSIAN J. OF INORGANIC CHEMISTRY, v. 9, no. 4, Apr. 1964.   p. 489-493.

POWELL, R.W. and R.P. TYE.   The Thermal and Electrical Conductivities of Zirconium and of Some Zirconium Alloys. J. OF LESS COMMON METALS, v. 3, 1961   p. 202-215.

POWELL, B.M. et al.   Lattice Dynamics of Niobium-Molybdenum Alloys.   PHYS. REV., v. 171, no. 3, July 15, 1968. p. 727-736.

RADHAKRISHNA, P. and M. NIELSEN.   Thermal Conductivity of Niobium-Zirconium Alloys at Low Temperatures.   PHYS. LETTERS, v. 6, no. 1, Aug. 15, 1963.   p. 36-38.

RADOSEVICH, L.G. and W.S. WILLIAMS.   Phonon Scattering by Conduction Electrons and by Lattice Vacancies in Carbides of the Transition Metals.   PHYS. REV., v. 181, no. 3, May 15, 1969.   p. 1110-1117.

RAETZ, K. and E. SAUR.   Study of Superconductivity in the Niobium-Aluminum System (In Ger.).   Z. FUER PHYSIK, v. 169, 1962.   p. 315-322.

RAIRDEN, J.R.   Thin Films of Niobium Nitride and Tantalum Nitride Deposited by Reactive Evaporation.   ELECTRO-CHEM. TECHNOLOGY, v. 6, no. 7-8, July-Aug. 1968.   p. 269-272.

RAKITIN, S.P. et al.   Some Results of the Use of Transition-Metal Carbides as Thermionic Emitters in Electronic Devices.   HIGH TEMP., v. 1, no. 1, July-Aug. 1963.   p. 124-127.

RALLS, K.M. et al.   High-Field Capabilities of High-Zirconium Niobium-Zirconium Superconducting Alloys.   In: CONF. FOR ADVANCED ELECTRONIC MATERIALS (AIME).   13 p.

RASSMANN, G. and A. MERZ.   Development in the Field of High Temperature Materials (In Ger.).   TECHNIK, v. 17, no. 2, Feb. 1962.   p. 74-79.

RAUCH, G.C. et al.   Ageing in Nb(Cb)-Ti-O Superconductors, with Appendix.   AIME METALL. SOC., TRANS., v. 242, no. 11, Nov. 1968.   p. 2263-2270.

REED, T.B. et al.   Superconducting Behavior of Some beta-Tungsten Structure Niobium Compounds and Their Alloys. In: METALLURGY OF ADVANCED ELECTRONIC MATERIALS, CONF. Philadelphia, Pa., Aug. 27-29, 1962.   New York, Interscience, 1963.   p. 71-87.

REED, T.B. et al.   Niobium Compounds with the Beta-Tungsten Structure.   M.I.T., Lincoln Laboratory.   Rept. No. 1. 1962.   AD 277 393.

REVOLINSKY, E. et al. Layer Structure Superconductor. SOLID STATE COMM., v. 4, no. 3, 1963. p. 59-61.

RINDERER, L. et al. Development and Superconductivity Properties of Niobium-Tin Diffusion Layers (In Ger.). ZEIT. FUER PHYSIK, v. 174, 1963. p. 405-422.

RIPLEY, R.L. The Preparation and Properties of Some Transition Phosphides. J. OF LESS COMMON METALS, v. 4, no. 6, Dec. 1962. p. 496-503.

RITTER, D.L. et al. The Niobium (Columbium)-Rhodium Binary System Part I: The Constitution Diagram. AIME METALL. SOC., TRANS., v. 230, no. 6, Oct. 1964. p. 1250-1259.

ROGENER, H. Superconductivity of Niobium Nitride (In Ger.). ZEIT. FUER PHYSIK, v. 132, no. 4, July 1952. p. 446-467.

ROGERS, B.A. and D.F. ATKINS. Zirconium-Columbium Diagram. J. OF METALS, v. 7, no. 9, Sept. 1955. p. 1034-1041.

ROSENBLUM, B. et al. Microwave Studies of Niobium Stannide. RCA REVIEW, v. 25, no. 3, Sept. 1964. p. 491-509.

ROTHWARF, F. et al. Superconducting Transition Temperatures and X-Ray Lattice Constants of Niobium Aluminum Antimony Alloys. AMERICAN PHYS. SOC., BULL., v. 7, no. 4, Ser. 2, 1962. p. 322.

RUZICKA, J. et al. Some Superconducting Properties of 25% Niobium-75% Zirconium Alloy. CZECH. J. OF PHYS., v. 16, no. 4, Ser. B, 1966. p. 338-341.

SADAGOPAN, V. and H.C. GATOS. Superconductivity in the Close-packed Intermediate Phases of the Vanadium-Iridium, Niobium-Iridium, Niobium-Rhodium, Tantalum-Rhodium, Niobium-Platinum, Tantalum-Platinum, and Other Related Systems. PHYS. STATUS SOLIDI, v. 13, no. 2, 1966. p. 423-427.

SADAGOPAN, V. et al. Fast Neutron Damage and the Current Carrying Capacity of Niobium Nitride. Z. FUER PHYSIK, v. 225, no. 3, 1969. p. 231-236.

SAHM, P.R. et al. Hot Pressing of Niobium Stannide and Columbium Stannide Composites and Some Resulting Superconductive Properties. AIME METALL. SOC., TRANS., v. 242, no. 4, Apr. 1968. p. 603-607.

SAINI, G.S. et al. Preparation and Characterization of Crystals of MX-and MX_2-Type Arsenides of Niobium and Tantalum. CANADIAN J. OF PHYS., v. 42, no. 3, Mar. 1964. p. 630-634.

SAITO, Y. et al. The Upper Critical Field of Niobium Nitride Film Prepared by Reactive Sputtering. APPLIED PHYS. LETTERS, v. 14, no. 9, May 1, 1969. p. 285-286.

SALTER, L.C. et al. Investigation of Current Degradation Phenomenon in Superconducting Solenoids. ATOMICS INTERNATIONAL, Canoga Park, Calif. Contract No. NAS 8-5356. Jan. 14, 1966. N66-23800.

SAMSONOV, G.V. and V.S. SINELNIKOVA. Study of Use of Chemical Bond in Aluminides of Several Transition Metals (In Russ.). AKAD. NAUK, SSSR. IZV. NEORGAN. MAT., v. 1, no. 7, 1965. p. 1071-1078.   A

SAMSONOV, G.V. and M.M. ANMONOVA. A Metastable Hydride Phase in the System (In Russ.). ZH. FIZ. KHIM., v. 35, no. 4, 1961. p. 900-904.

SAMSONOV, G.V. and K.I. PORTNOY. Alloys Based on High-Melting Compounds. Rept. No. FTD-TT-62-430/1+2. July 24, 1962. 387 p. AD 283 859.

SAMSONOV, G.V. and V.G. GREBENKINA. Temperature Coefficient of Electroresistance of Some High-Melting Compounds. SOVIET POWDER METALL. AND METAL CERAM., no. 2, Feb. 1968. p. 107-111.

SAMSONOV, G.V. and V.S. SINELNIKOVA. The Resistivity of Refractory Compounds at High Temperatures. SOVIET POWDER METALL. AND METAL CERAM., no. 4, July-Aug. 1962. p. 272-274. B

SAMSONOV, G.V. et al. Certain Electrophysical Properties of Disilicides of Transition Metals in Periodic Groups V and VI. SOVIET POWDER METALL. AND METAL CERAM., no. 4, Apr. 1969. p. 292-295. A

SAMSONOV, G.V. and V.N. PADERNO. Preparation of Carbide Mixed Crystals and Study of their Physical Properties (In Ger.). PLANNSEEBERICHTE FUER PULVERMETALLURGIE, v. 12, no. 1, 1964. p. 19-30.

SAMSONOV, G.V. et al. Electrical and Physical Properties of Defined Compounds and the Homogeneity Domains of Transition Metal Carbides of Groups 4 and 5 in the Periodic Table to Temperatures of 2500°C (In Fr.). REV. HAUTES TEMP. REFRACTAIRES, v. 3, no. 2, Apr.-June 1966. p. 179-184.   B

SAMSONOV, G.V. and A.D. PANASYUK. Some Electrophysical Properties of Niobium and Zirconium Carbides in their Homogeneous Regions. HIGH TEMP., v. 4, no. 2, Mar.-Apr. 1966. p. 203-209.

SAMSONOV, G.V. et al. Emission Coefficient of High-Melting Compounds. SOVIET POWDER METALL. AND METAL CERAM., no. 5, May 1969. p. 374-379.   C

SCHINDLER, H.C. and F.R. NYMAN. Electromagnetic Performance of Niobium-Stannide Ribbon. RCA REVIEW, v. 25, no. 3, Sept. 1964. p. 570-581.

SCHÖNBERG, N. Some Features of the Nb-N and Nb-N-O Systems. ACTA CHEM. SCANDINAVICA, no. 8, 1954. p. 208-212.

SCHRÖDER, E. Superconductive Compounds of Niobium (In Ger.). Z. FUER NATURFORSCH., v. 12a, no. 3, Mar. 1957. p. 247-256.

SCOTT, B.A. et al. Magnetic Susceptibility and Nuclear Magnetic Resonance Studies of Transition-Metal Monophosphides. J. OF CHEM. PHYS., v. 48, no. 1, Jan. 1, 1968. p. 263-272.

SEKULA, S.T. et al. Longitudinal Critical Currents in Cold-Drawn Superconducting Alloys. In: OAK RIDGE NAT. LAB., SOLID STATE DIV. Annual Progress Report May 31, 1963. p. 41-42.   A

SEKULA, S.T. et al. Dependence of Superconducting Critical Current of Niobium-Zirconium Alloys on Magnetic-Field Orientation. In: OAK RIDGE NAT. LAB., SOLID STATE DIV. Annual Progress Report for the period ending Aug. 31, 1962. ORNL-3364. p. 52-56.   B

SELTE, K. and A. KJEKSHUS. On the Magnetic Properties of Niobium Selenides and Tellurides. ACTA CHEM. SCANDINAVICA, v. 19, no. 1, 1965. p. 258-260.

SHAHEEN, L.C. Superconductive Columbium-Tin Wire. ELECTRO-TECHNOLOGY, v. 72, no. 5, Nov. 1963. p. 11-12.

SHAPIRA, Y. and L.J. NEURINGER. Magnetoacoustic Attenuation in High-Field Superconductors. PHYS. REV., v. 154, no. 2, Feb. 10, 1967. p. 375-385.   A

SHAPIRA, Y. and L.J. NEURINGER. Upper Critical Fields of Niobium-Titanium Alloys: Evidence for the Influence of Pauli Paramagnetism. PHYS. REV., v. 140, no. 5A, Nov. 29, 1965. p. 1638-1644.   B

SHUKOVSKY, H.B. et al. The Effect of Heat Treatment on the Microstructure and Superconducting Properties of a 45% Niobium-55% Zirconium Alloy. AIME METALL. SOC., TRANS., v. 233, no. 10, Oct. 1965. p. 1825-1832.

SHULISHOVA, O.I. Superconductivity and the Nature of the Bond in Carbides and Nitrides of Transition Metals and their Solid Solutions with Sodium Chloride-Type Structure. Rept. No. LA-TR-68-31. Aug. 1968. 27 p. N69-13266.

SINELNIKOVA, V.S. and A.S. GORALNIK. Electronic Nature of Aluminides. Communication II. SOVIET POWDER METALL. AND METAL CERAM., no. 1, Jan. 1969. p. 58-64.

SINELNIKOVA, V.S. Physical Properties of Aluminides of Transition Metals (In Russ.). POROSHKOVAYA MET., no. 6, 1966. p. 64-67.

SIROTA, N.N. and E.A. OVSEICHUK. Superconducting Properties of Vanadium-Niobium Alloys. SOVIET PHYS. DOKL., v. 12, no. 5, Nov. 1967. p. 516-517.   B

SIROTA, N.N. and E.A. OVSEICHUK. Superconducting Properties of Vanadium and Niobium Alloys (In Russ.). AKAD. NAUK BELARUS. SSR., VESTI, SER. FIZ.-MAT., v. 1, 1967. p. 131.   A

SOUSA, J.B. Lattice Thermal Conductivity of Tantalum-Niobium and Niobium-Molybdenum Solid Solution Alloys in Normal and Superconducting States. J. OF PHYS., C, Ser. 2, v. 2, no. 4, Apr. 1969. p. 629-639.

STAUFFER, R.A. et al. A New Form of Niobium Stannide. SOLID STATE DESIGN, v. 5, no. 5, May 1964. p. 11-13.

STORMS, E.K. and N.H. KRIKORIAN. The Variation of Lattice Parameter with Carbon Content of Niobium Carbide. J. OF PHYS. CHEM., v. 63, no. 10, Oct. 1959. p. 1747-1749. [A]

STORMS, E.K. and N.H. KRIKORIAN. The Niobium-Niobium Carbide System. J. OF PHYS. CHEM., v. 64, no. 10, Oct. 1960. p. 1471-1477. [B]

SWARTZ, P.S. and C.H. ROSNER. Characteristics of a New Application of High-Field Superconductors. J. OF APPLIED PHYS., v. 33, no. 7, July 1962. p. 2292-2300.

SWARTZ, P.S. et al. Effect of Fast-Neutron Irradiation on Magnetic Properties and Critical Temperature of Some Type II Superconductors. APPLIED PHYS. LETTERS, v. 4, no. 4, Feb. 15, 1964. p. 71-73.

SUKHAREVSKII, B.Ya. et al. Some Features of the Temperature Dependence of the Specific Heat of an Niobium-Titanium Alloy at the Transition to the Superconducting State. SOVIET PHYS. JETP, v. 27, no. 6, Dec. 1968. p. 897-899.

TANIGUCHI, S. et al. The Magnetic Susceptibilities of Some Transition Metal Alloys and the Corresponding Density of States Curves. ROYAL SOC., PROC., v. 265, no. 1323, Feb. 6, 1962. p. 502-518.

TAYLOR, W. et al. Solid Solubility Limits of Yttrium and Scandium in the Elements Tungsten, Tantalum, Molybdenum, Niobium, and Chromium. J. OF LESS COMMON METALS, v. 9, no. 3, Sept. 1965. p. 214-232.

TESTARDI, L.R. et al. Lattice Instability of High-Transition-Temperature Superconductors. I. Polycrystalline A-15 Compounds. PHYS. REV., v. 154, no. 2, Feb. 10, 1967. p. 399-401.

THOMAS, L.K. Optical Constants of Tantalum-Tungsten- and Niobium-Molybdenum-Alloys at Incandescent Temperatures (In Ger.). Z. FUER ANGEWAND. PHYS., v. 27, no. 3, July 1969. p. 209-213.   A

THOMAS, L.K. Normal Spectral Emissivity of Tantalum-Tungsten and Niobium-Molybdenum Alloys. J. OF APPLIED PHYS., v. 39, no. 8, July 1968. p. 3737-3742.   B

TOTH, L.E. et al. Low Temperature Heat Capacities of Superconducting Niobium and Tantalum Carbides. ACTA METALLURGICA, v. 16, no. 9, Sept. 1968. p. 1183-1187.

TOULOUKIAN, Y.S. Thermophysical Properties of High Temperatures Solid Materials. In: MacMILLAN COMPANY, NEW YORK COLLIER-MacMILLAN LIMITED, London. V. 5, 1967.   [TPRC]

TREUTING, R.G. et al. Effect of Heat-Treatment on Niobium-Zirconium Superconducting Alloys. In: HIGH MAGNETIC FIELDS. Edited by H. Kolm, et al. PROC. OF THE INTERNAT. CONF. ON HIGH MAGNETIC FIELDS, held at MIT Nov. 1-4, 1961. MIT and Wiley, c1962. p. 597-601.

TYAN, Y.S. et al. Low Temperature Specific Heat Study of the Electron Transfer Theory in Refractory Metal Borides. J. OF PHYS. AND CHEM. OF SOLIDS, v. 30, no. 4, Apr. 1969. p. 785-792.

UKEI, K. and E. KANDA. Superconductive Anomaly in Specific Heats of Some Niobium and Vanadium Compounds. In: LOW TEMP. CALORIMETRY CONF., 1966 PROC., held at Helsinki, Finland, Aug. 26-29. Ser. A, VI Physics. Edited by O.V. Lounasmaa. p. 104-107.

ULLMAIER, H.A. AC Measurements on Hard Superconductors. PHYS. STATUS SOLIDI, v. 17, no. 2, 1966. p. 631-643.

ULYANOV, R.A. and N.D. TARASOV. Certain Regularities in the Variation in the Properties of Alloys on Niobium Base. PHYS. OF METALS AND METALL., v. 17, no. 2, 1964. p. 60-64.

VAN MAAREN, M.H. and H.B. HARLAND. An Energy Band Model of Niobium- and Tantalum-Dichalcogenide Superconductors. PHYS. LETTERS, v. 29A, no. 9, July 28, 1969. p. 571-573.

VAN MAAREN, M.H. and G.M. SCHAEFFER. Superconductivity in Group Va Dichalcogenides. PHYS. LETTERS, v. 20, no. 2, Feb. 1, 1966. p. 131.

VAN OOIJEN, D.J. and A.S. VAN DER GOOT. Critical Currents of Superconducting Niobium-Oxygen Alloys. PHILIPS RES. REPTS., v. 20, no. 2, Apr. 1965. p. 162-169.

VAN OOIJEN, D.J. et al. Superconductivity of 33% Niobium-67% Tin. PHYS. LETTERS, v. 3, no. 3, Dec. 15, 1962. p. 128-129.

VAN OSTENBURG, D.O. et al. NMR, Magnetic Susceptibility and Electronic Specific Heat of Niobium and Molybdenum Metals and Niobium-Technetium and Niobium-Molybdenum Alloys. PHYS. SOC. OF JAPAN, J., v. 18, no. 12, Dec. 1963. p. 1744-1754

VAN VUCHT, J.H.N. et al. Some Investigations on the Niobium-Tin Phase Diagram. PHILIPS RES. REPTS., v. 20, no. 2, Apr. 1965. p. 136-161.

VEAL, B.W. et al. The Heat Capacity of Niobium-Molybdenum Alloys. In: LOW TEMP. CALORIMETRY CONF., 1966 PROC., held at Helsinki, Finland, Aug. 26-29. Ser. A, VI Physics. Edited by O.V. Lounasmaa. p. 108-114.

VETRANO, J.B. and R.W. BOOM. High Critical Current Superconducting Titanium-Niobium Alloy. J. OF APPLIED PHYS., v. 36, no. 3, Part 2, Mar. 1965. p. 1179-1180.

VIELAND, L.J. High-Temperature Phase Equilibrium and Superconductivity in the System Niobium-Tin. RCA REVIEW, v. 25, no. 3, Sept. 1964. p. 366-378.

VIELAND, L.J. and A.W. WICKLUND. Specific Heat of Niobium-Tin. PHYS. REV., v. 166, no. 2, Feb. 10, 1968. p. 424-431.

VIELAND, L.J. et al. Study of Transition Temperatures in Superconductors. Final Rept. Mar. 11, 1968-Mar. 10, 1970. Contract No. NAS 8-21384. Mar. 10, 1970. 122 p. N70-30793-804.

VINEN, W.F. and A.C. WARREN. Flux Flow Resistivity in Type II Superconductors I. Experimental Results. PHYS. SOC., PROC., v. 91, Pt. 2, no. 572, June 1967. p. 399-408.

WALKER, M.S. and M.J. FRASER. Field - Dependent Anisotropy of the Critical Current in Niobium-Zirconium Rolled Strip. In: SUPERCONDUCTORS. Edited by M. Tanenbaum and W.V. Wright. New York, Interscience, Feb. 18, 1962. p. 99-113.

WEINBERG, I. and C.W. SCHULTZ. Thermoelectric Power in Niobium-Zirconium Alloys. J. OF PHYS. AND CHEM. OF SOLIDS, v. 27, no. 2, Feb. 1966. p. 474-476.

WERNICK, J.H. et al. Evidence for a Critical Magnetic Field in Excess of 500 Kilogauss in the Superconducting Vanadium-Gallium System. In: HIGH MAGNETIC FIELDS. Edited by H. Kolm, et al. PROC. OF THE INTERNAT. CONF., ON HIGH MAGNETIC FIELDS, held at MIT Nov. 1-4, 1961. MIT and Wiley, c1962. Chapter 75, p. 609-614.

WIEDERMANN, W. Electrical and Magnetic Properties of Niobium Stannide in the Superconductive Transition Range (In Ger.). ZEIT. FUER PHYSIK, v. 151, no. 3, May 1958. p. 307-327.

WILHELM, H.A. et al. Columbium-Vanadium Alloy System. AIME METALL. TRANS., J. OF METALS, v. 6, no. 8, Aug. 1954. p. 915-918.

WILLENS, R.H. et al. Superconductivity of 75% Niobium-25% Aluminum. SOLID STATE COMM., v. 7, no. 11, June 1969. p. 837-841.

WILLIAMS, D.E. and W.H. PECHIN. The Tantalum-Columbium Alloy System. TRANSACTIONS OF THE ASM, v. 50, 1958. p. 1081-1089.

WONG, J. Metallurgical Aspects of Superconducting Niobium-Zirconium Alloys. In: SUPERCONDUCTORS. Edited by M. Tanenbaum and W.V. Wright. New York, Interscience, Feb. 18, 1962. p. 83-97.

WOOD, E.A. and B.T. MATTHIAS. The Crystal Structures of 75% Niobium-25% Gold and 75% Vanadium-25% Gold. ACTA CRYST., no. 9, 1956. p. 534.

WOOD, E.A. et al. Beta-Wolfram Structure of Compounds Between Transition Elements and Aluminum, Gallium and Antimony. ACTA CRYST., v. 11, 1958. p. 604-606.

WOODARD, D.W. and G.D. CODY. Anomalous Resistivity of Niobium Stannide. RCA REVIEW, v. 25, no. 3, Sept. 1964. p. 393-404.

ZAR, J.L. Alternating Current Resistance of Nonideal Superconductors. J. OF APPLIED PHYS., v. 35, no. 5, May 1964. p. 1610-1615.

ZEGLER, S.T. Superconductivity in 75% Chromium-25% Silicon-Type Ternary Phases with Niobium and Group VIII Metals. PHYS. REV., v. 137, no. 5A, Mar. 1, 1965. p. A1438-A1440.

SUPPLEMENTARY BIBLIOGRAPHY

BACHMANN, R. et al. Optical Properties and Superconductivity of $NbSe_2$. SOLID STATE COMM., v. 9, no. 1, 1971. p. 57-60

FINK, H.J. et al. High Field Superconductivity of Carbides. PHYS. REV., v. 138, no. 4A, May 1965. p. 1170-1173

FRAAS, L.M. et al. Photoluminescence Studies of Superconducting Nb_3Sn. SOLID STATE COMM., v. 8, no. 24, Dec. 1970. p. 2113-2115.

GALASSO, F. and J. PYLE. Nb_3Si, a Superconductor with the Ordered Cu_3Au Structure. ACTA CRYST., v. 16, 1963. p. 228-229.

KESKAR, K.S. et al. Superconducting Transition Temperatures of rf Sputtered NbN Films. JAPANESE J. OF APPLIED PHYS., v. 10, no. 3, Mar. 1971. p. 370-374.

TERAO, N. New Phases of Niobium Nitride. J. OF THE LESS-COMMON METALS, v. 23, no. 2, Feb. 1971. p. 159-169.